# From Rio+20 to a New Development Agenda

Twenty years after the United Nations Conference on Environment and Development, "The Earth Summit," the Rio+20 conference in 2012 brought life back to sustainable development by putting it at the centre of a new global development partnership, one in which sustainable development is the basis for eradicating poverty, upholding human development and transforming economies.

Written by practitioners and participants involved in the multilateral process of negotiations, this book presents a unique insider analysis of not only what happened and why, but also where the outcomes might impact in the future, particularly in the United Nations' development agenda beyond 2015.

The book throws light on the changing nature of multilateralism and questions frequent assumptions on how policy is defined within the United Nations. It shows that Rio+20 was more than an international meeting; it represented a culminating point of decades of successes and failures and a watershed moment for seminal concepts, ideas and partnerships, including the green economy, zero tolerance on land degradation, the introduction of Sustainable Development Goals, the creation of national measurements of consumption, production and well-being that are intended to go beyond GDP, the introduction of national green accounting and the commitment of billions of dollars for sustainable development partnerships, including Sustainable Energy for All.

The authors conclude by mapping out a new agenda for development in 2015, when the current Millennium Development Goals framework is due to expire – an agenda that will restore faith in the United Nations and inspire a global response to the demographic, economic and environmental challenges that will define our future in the decades to come.

**Felix Dodds** is former Executive Director of the Stakeholder Forum for a Sustainable Future, an Associate Fellow at the Tellus Institute, and on the board of advisers for the Collaborative Institute for Oceans, Climate and Security, Asia European Foundation ENVforum Steering Committee and the Great Transition.

**Jorge Laguna-Celis** is Senior Advisor on sustainable development to the President of the UN General Assembly; prior to that, he was a negotiator and expert on international environmental governance and sustainable development. He has been a member of the Mexican Foreign Service since 2006 and is based in New York.

**Liz Thompson** is a former Minister of Energy and Environment of Barbados. She was honoured by UNEP in 2008 with the prestigious "Champion of the Earth" award for her work in climate change and environment. She was one of two executive coordinators appointed by the United Nations at the level of Assistant Secretary-General to support the Rio+20 process and negotiations.

# From Rio+20 to a New Development Agenda

## Building a bridge to a sustainable future

Felix Dodds, Jorge Laguna-Celis and Elizabeth Thompson

Routledge
Taylor & Francis Group
LONDON AND NEW YORK

earthscan
from Routledge

First published 2014
by Routledge
4 Park Square, Milton Park, Abingdon, Oxon OX14 4RN
605 Third Avenue, New York, NY 10017

*Routledge is an imprint of the Taylor & Francis Group, an informa business*



*British Library Cataloguing-in-Publication Data*
A catalogue record for this book is available from the British Library

*Library of Congress Cataloging-in-Publication Data*
Dodds, Felix.
From Rio+20 to a new development agenda : building a bridge to a sustainable future / Felix Dodds, Jorge Laguna Celis, Elizabeth Thompson.
pages cm
1. Rio+20 (Conference) (2012 : Rio de Janeiro, Brazil)–Influence.
2. Sustainable development–History–21st century. 3. Renewable energy sources. 4. Environmental protection. 5. Environmental policy.
6. Economic policy. I. Laguna Celis, Jorge. II. Thompson, Elizabeth, 1955–
III. Title.
HC79.E5D633 2014
338.9'27–dc23 2013033586

ISBN: 978-0-415-71653-6 (hbk)
ISBN: 978-0-415-71654-3 (pbk)
ISBN: 978-1-315-87986-4 (ebk)

Typeset in Bembo
by Keystroke, Station Road, Codsall, Wolverhampton

Disclaimer: Mr Jorge Laguna-Celis prepared this book in his personal capacity and received no remuneration or honorarium associated with its preparation. The views expressed do not necessarily reflect the views of the Office of the President of the 68th Session of the General Assembly, the United Nations or the Permanent Mission of Mexico to the United Nations.

Felix Dodds
To my amazing children, Merri and Robin, who I hope will live and laugh in a more sustainable world

Jorge Laguna-Celis
To Octavio Olin, who will turn 40 in 2050

Elizabeth Thompson
To my parents, Harry and Pauline, born in 1923 and 1924, who bridge the past, live the present, and envision a sustainable future

Multilateral negotiations are very complex. Issues such as peace and security, sustainable development and climate change have all proven how difficult it is to negotiate to consensus amongst the global family of nations. The Rio+20 Conference on Sustainable Development ushered in new policy and institutional approaches for achieving sustainable development. The three authors of this book represent the various stakeholders in the multilateral negotiation process – countries, civil society and the UN system. The authors are uniquely positioned to offer readers a better insight into how a consensus in multilateral negotiations emerge, the result of Rio+20 and the evolving global development agenda.

**H.E. Dr. John W. Ashe** (Co-Chairman, Rio+20 Preparatory Committee, Permanent Representative of Antigua and Barbuda to the United Nations)

Long-term sustainability solutions can only be brokered at the international level – and a huge amount of energy and expertise has gone into those international agendas since 1972. Through the Stakeholder Forum, Felix Dodds has been at the heart of that global diplomacy for more than 20 years, allowing him and his co-authors, Liz Thompson and Jorge Laguna-Celis to draw on that wealth of experience and theirs in order to map out a robust and inspiring road map for a better world.

**Jonathon Porritt** (Founder Director, Forum for the Future and former Chair of Stakeholder Forum)

This is the ultimate "insider's guide" to the Rio+20 process, taking the reader from the preliminary discussions back in 2007 all the way through the conference itself and into the future. In tabloid terminology, Felix Dodds, Jorge Laguna-Celis and Liz Thompson give us "the thrills, the spills, the personalities and power-plays" that shaped the summit and its diverse outcome. Whether you regard the outcome as primarily a missed opportunity due to its fundamental incompatibility with respect for planetary boundaries and its failure to match President Lula's pre-conference vision of an event that would lead to "a new and more balanced distribution of wealth" in the world, as I continue to do, or whether you share the authors' optimism regarding the re-moulding of international institutions and the voluntary commitments secured in Rio, this book will give you a deeper and broader understanding of why Rio reached the outcomes it did. And as it makes clear, there is still much to play for, particularly in regard of the sustainable development goals, which could yet provide impetus for a global move towards the future we both want and need.

**Richard Black** (former BBC Environment Correspondent)

The Earth Summit of 1992 was a pivotal event in multilateralism; it set the world on a journey to sustainable development. In 2012 the world's governments returned to Rio to rechart that path in the light of past successes and failures and the new challenges which lay ahead. In this book, three development practitioners conduct an interdisciplinary assessment of the outcome document of the Rio+20 Conference on Sustainable Development, "The Future We Want" and the post-2015 development agenda. It is a refreshing and timely contribution to the body of development literature.

**Sir Shridath Ramphal** (Former Commonwealth Secretary General)

This book is a decisive contribution to a better understanding of the Rio + 20 negotiations which in many ways were the genesis of the Post-2015 process. The negotiations were dauntingly complex given the range of issues covered, and this book is both timely and seminal as it will enable those who did not participate in the process, for the first time, to fully appreciate the scope of the negotiations. This will be a vital referent for the discussions under the Post-2015 process which need to be widely participatory. This book will contribute to facilitating informed and substantive participation by a wide range of stakeholders.

**Paula Caballero Gomez** (Director, Ministry of Foreign Affairs, Colombia)

The landscape of geopolitics is rapidly changing, as is our view of how we can live and prosper within the planet's boundaries while attaining sustainable global development. Dodds, Laguna-Celis and Thompson have captured the nuances of these changes and challenges, presenting them in a way that enhances our understanding of the major development issues of our time and the response of the multilateral system to those issues.

**Dame Billie Miller** (Former Deputy Prime Minister, Minister of Foreign Affairs and Trade of Barbados)

Out of respect for our children we as politicians must show will and courage to embark on a new course towards turning the unsustainable use of natural resources around. It is my hope we will succeed in setting ambitious global sustainability goals in the UN in 2015 to continue the hard work done at Rio+20. This book highlights that it is not about changing gears. It is about changing lanes.

**Ida Auken** (Minister of Environment, Denmark)

This is an invaluable contribution to the on-going conversation on the future of our planets and its present and future inhabitants. The ideas for action are anchored in an erudite appraisal of the results of the Rio+20 summit and deserve careful consideration by policy makers, advocacy groups and concerned citizens.

**Shafqat Kakakhel** (former Deputy Director of UNEP)

This book is an animated, engaging and well documented tour through the recent history of development, culminating in Rio+20 and its declaration. But it goes beyond history and points a path for achieving the kind of development that human beings long for and how we must plan for it. This is probably the greatest challenge of our time and this book gives us hope that we are up to it.

**Sir (Dr) George Alleyne** (Director Emeritus, PanAmerican Health Organization, Chancellor of the University of the West Indies)

There are many who are only too happy to dismiss multilateral negotiations because of their painfully slow progress and bizarre esoteric practices. Sections of the media feel cliché bound to portray them as a junket circuit for politicians and officials, and delight in calling them a waste of time and money. And it is undoubtedly the case that drive, passion, ideas and excitement are much more obvious in the myriad of grassroots people's movements around the world campaigning for change; in the dynamic partnerships between forward thinking cities and regions; and in the innovative approaches that thousands of communities and some leading companies are taking to address the challenges and opportunities of sustainable development. But if these thousands of actions aiming to create a better world are to come together in a way that enables humanity to prosper fairly within environmental limits, we need to keep our faith in a vision of multilateralism where the world's countries, slowly but surely, build the common frameworks that are needed to deliver genuine sustainable development. This book does an excellent job of documenting the journey to the Rio+20, the details and outcomes of the summit negotiations, and critically what needs to happen in the years ahead. It makes sense of where we have got to on this long journey, and where we need to go next. Nothing could be more important.

**Craig Bennett** (Director of Policy and Campaigns, Friends of the Earth (England, Wales and Northern Ireland))

# Contents

# List of illustrations

## Figures

## Boxes

# Foreword

Much has been written about the pathway to a sustainable future through a new development agenda but this is the most thorough and authoritative account. It is unique in the quality of the research on which it draws and the insights of some of the leading professionals in the field. Indeed, there is no better source of well researched and reliable information and analysis than this book presents. It is derived from the experience and knowledge of the leading professionals in the field. This makes it the best source of reference available on the wide range of subjects which it covers and the systemic relationship that enables them to be seen as a whole.

It is particularly useful for me in my own continuing work in this area. I have found this comprehensive report indispensable as my main bible in the field.

The unsustainable nature of our current economic system was dramatically revealed by both the climate change and the environmental economic crises. They are inextricably linked on a systemic, integrated basis and cannot be managed as separate and competing issues.

The climate change challenge requires us to make changes in the fundamental nature and functioning of our economic system and resist the temptation merely to patch up the existing system to enable it to continue, however temporarily, on the pathway that led to its crisis. Only through fundamental change can we transcend these crises and rebuild the economic and social foundations of our civilization to ensure its survival and sustainability.

Throughout human history, civilizations have risen and fallen, often due to mismanagement of their environment and the resources on which they depended. The consequences were clearly devastating for those affected, but there was always somewhere else for them to go. The climate change crisis is fundamentally different because it is global in scale and affects the survival and sustainability of all nations and people. It is also different in that we know its causes and probable consequences.

We are the first generation ever to have the responsibility for our own future. What we do, or fail to do, will determine the future of life on Earth. This requires unprecedented levels of cooperation both within and amongst nations. But it does not require homogeneity in our life styles or cultures. We can learn from nature that the healthiest and more sustainable natural ecological systems are those which maintain the highest degree of diversity and variety.

The fundamental changes I believe must take place at the level of individual people as well as nations promise to produce improved conditions of life and a more secure and sustainable future for all people. It is instructive to reflect that the conditions required for sustaining life as we know it have only existed for a small portion of the Earth's history and within relatively narrow parameters. Our very existence is now at risk and its future is literally in our own hands.

The good news so well recited here is that we have the knowledge and capacity to ensure our survival. In the more than forty years since the United Nations Conference on the Human Environment in Stockholm, Sweden, in 1972 put the environment issue on the global agenda, the concept of the environment has been broadened to encompass the wide range of economic, social, biological and climatic conditions that affect the quality and the sustainability of life on Earth. It is this larger contest that is so well presented in this impressive book.

In recent times, there has been a great expansion of environmental education and research and many universities and institutes now have extensive programs in addressing these issues. There is no better compendium of information and guidance on the complex of issues that constitutes sustainable development. The authors have done a great service in producing this impressive work.

*Maurice Strong was secretary-general of the Stockholm UN Conference on Human Settlement (1972) and was secretary-general of the UN Conference on Environment and Development (1992). He was also the first executive director of UNEP.*

# Author biographies

**Felix Dodds** is a writer, activist and futurist who has been active at the global level on sustainable development agenda for more than 20 years. He is currently co-director of the 2014 Nexus Conference on climate, energy, food and water as well as being a Fellow at the Global Research Institute at the University of North Carolina and an Associate Fellow at the Tellus Institute in Boston. He was executive director of Stakeholder Forum for a Sustainable Future for 20 years (1993–2012) and co-director of the Water and Climate Coalition from 2007 to 2012. He chaired the United Nations DPI 64th NGO Conference "Sustainable Societies – Responsive Citizens" (2011).

Felix co-chaired the NGO Coalition at the United Nations from 1997 to 2001 and is credited with introducing the concept of stakeholder dialogues to the United Nations Commission on Sustainable Development in 1996.

This is his tenth book on sustainable development. His last one, *Only One Earth: The Long Road via Rio to Sustainable Development*, was co-written with Michael Strauss and Maurice Strong.

**Jorge Laguna-Celis** is a negotiator, diplomat and writer. Since 2007 he has worked for a more effective, responsive and representative multilateral system. He is currently the Senior Advisor on sustainable development to the President of the United Nations General Assembly.

As lead negotiator for his country, Mexico, during the Rio+20 Conference and chair of the historic UN General Assembly resolution on the strengthening of the environmental pillar of sustainable development (67/213 of December 2012), he contributed to the reform of international environmental governance and to the establishment of the United Nations Environment Assembly.

He has also built international agreements in the fields of adaptation, resiliency and sustainable urbanization leading to the Third World Conference on Disaster Risk Reduction and to the Third United Nations Conference on Sustainable Urban Development. He has served as deputy head of the Mexican Embassy in Nairobi and deputy permanent representative to the United Nations Environment Programme (UNEP), and to the United Nations Human Settlements Programme (UN-HABITAT) as well as international consultant for the Organisation for Economic Co-operation and Development (OECD) in Paris.

As an author he has published regular columns and numerous publications on topics ranging from international affairs and development, political and intellectual history of Spain to international environmental governance. He holds a BA in International Relations from El Colegio de México and a MA in International Trade from the Institut d'Etudes Politiques, Paris. Since 2006 he has been a member of the Mexican Foreign Service.

**Liz Thompson** is an international consultant working in the areas of sustainable development, energy, environment and Small Island Developing States. In 2010 she was appointed as a UN Assistant Secretary-General with responsibility as one of two executive coordinators of the Rio+20 Conference. In this role she led the creation and development of Rio+20's most successful multi-stakeholder initiative, the Higher Education Sustainability Initiative (HESI).

She was elected to the Parliament of Barbados in 1994, when she was also appointed as a government minister, serving in both these roles until 2008. In January of that year she was appointed as leader of Opposition/Minority business in the Barbados Senate. During her political career she held several portfolios, including those of energy and the environment. In 2008 she was awarded the United Nations Environment Programme's prestigious Champion of the Earth Award for her work in the environment, sustainable development and climate change. Her 14 years as a minister resulted in her involvement in many multilateral conferences and processes.

As a minister she led Barbados's national sustainable development policy and strategies, including the development of Barbados's first green buildings, its first sustainable energy policy intended to transition the country to 30 per cent renewables, the development of Latin America and the Caribbean's first national green economy policy (2007), and numerous ground-breaking environmental policies and initiatives. Working under contract to the Organization of American States, she led the teams that developed sustainable energy policies for Grenada and Dominica.

Her academic qualifications are as an attorney at law, and she is also trained in commercial arbitration, petroleum negotiations and alternative dispute resolution. She holds an LLB from the University of the West Indies, an LEC from the Hugh Wooding Law School, an LLM from the Robert Gordon University, Scotland, and an MBA with distinction from the University of Liverpool. She has lectured and written on her areas of work and interest.

# Acronyms and abbreviations

| | |
|---|---|
| ALBA | Alianza Bolivariana para los Pueblos de Nuestra América (Bolivarian Alliance for the Peoples of Our America) |
| AMR | Annual Ministerial Review (ECOSOC) |
| ANPED | Northern Alliance for Sustainability |
| AOSIS | Alliance of Small Island States |
| ASEAN | Association of Southeast Asian Nations |
| BASIC | Brazil, South Africa, India and China |
| BIMUN | Bonn International Model United Nations |
| BMZ | Federal Ministry for Economic Cooperation and Development (Germany) |
| BOLT | Build Operate Lease Transfer |
| BOOT | Build Own Operate Transfer |
| BPOA | Barbados Programme of Action |
| BRICS | Brazil, Russia, India, China, South Africa |
| CARICOM | Caribbean Community |
| CELAC | Community of Latin American and Caribbean States |
| CEO | chief executive officer |
| CFSD | Committee on Financing Sustainable Development |
| COP | Conference of Parties (of the UN FCCC) |
| CPR | Committee of Permanent Representatives |
| CSD | Commission on Sustainable Development (United Nations) |
| CSR | corporate social responsibility |
| CSRC | Corporate Sustainability Reporting Coalition |
| CSW | Commission on the Status of Women (United Nations) |
| DCF | Development Cooperation Forum (ECOSOC) |
| DESA | *see* UN DESA |
| DPI | Department of Public Information (United Nations) |
| EAC | East African Community |
| ECLAC | Economic Council for Latin America and the Caribbean (United Nations) |
| ECOSOC | Economic and Social Council (United Nations) |
| EU | European Union |

| | |
|---|---|
| FCCC | *see* UNFCCC |
| FTAA | Free Trade Area of the Americas |
| FTT | financial transaction tax |
| G5 | group comprising Brazil, China, India, Mexico and South Africa |
| G7 | group of seven industrialized countries (Canada, France, Germany, Italy, Japan, the United Kingdom, the United States of America) |
| G8 | G7 plus Russia |
| G20 | Argentina, Australia, Brazil, Canada, China, France, Germany, India, Indonesia, Italy, Japan, the Republic of Korea, Mexico, Russia, Saudi Arabia, South Africa, Turkey, the United Kingdom, the United States of America plus the European Union |
| G77 | group of 77 developing countries |
| GCF | Green Climate Fund |
| GDP | gross domestic product |
| GEG | Global Environmental Goal |
| *GEO* | *Global Environmental Outlook* |
| GIZ | German Society for International Cooperation |
| GMEF | Global Ministerial Environmental Forum (UNEP) |
| GNP | gross national product |
| GPO | Global Partnership on Oceans |
| GRI | Global Reporting Initiative |
| GSP | Global Sustainability Panel |
| HDI | Human Development Index |
| HESI | Higher Education Sustainability Initiative |
| ICC | International Chamber of Commerce |
| ICLEI | International Council for Local Environmental Initiatives |
| ICSU | International Council for Science |
| IDB | Inter-American Development Bank |
| IEEE | Institute of Electrical and Electronics Engineers |
| IEG | international environmental governance |
| IFF | Intergovernmental Forum on Forests (United Nations) |
| IFSD | institutional framework for sustainable development |
| IGBP | International Geosphere-Biosphere Programme |
| IGO | intergovernmental organization |
| IHDP | International Human Dimensions Programme on Global Environmental Change |
| IISD | International Institute for Sustainable Development |
| ILO | International Labour Organization |
| IMF | International Monetary Fund |
| IPCC | Intergovernmental Panel on Climate Change |
| IRENA | International Renewable Energy Agency |

| | |
|---|---|
| IUCN | International Union for Conservation of Nature |
| ISSC | International Social Science Council |
| JPoI | Johannesburg Plan of Implementation |
| LAMG | Local Authorities Major Group |
| LDCs | least developed countries |
| LPG | liquefied petroleum gas |
| MDG | Millennium Development Goal |
| MEA | multilateral environmental agreement |
| MOI | means of implementation |
| MRV | Monitoring, Reporting and Verification |
| MSI | Mauritius Strategy of Implementation |
| NCSD | National Council for Sustainable Development |
| NGO | non-governmental organization |
| NRDC | Natural Resources Defense Council |
| nrg4SD | Network for Regional Government for Sustainable Development |
| NSDS | National Sustainable Development Strategy |
| OECD | Organisation for Economic Co-operation and Development |
| OECS | Organization of Eastern Caribbean States |
| OESC | UNDESA Office for ECOSOC Support and Coordination |
| ODA | official development assistance |
| OFID | OPEC Fund for International Development |
| OPEC | Organization of the Petroleum Exporting Countries |
| OWG | Open Working Group on Sustainable Development Goals |
| PIC | prior informed consent |
| PrepCom | Preparatory Committee (for UN conferences and summits |
| SADC | Southern African Development Community |
| SCP | sustainable consumption and production |
| SDG | Sustainable Development Goal |
| SDPI | Sustainable Development Policy Institute |
| SDSN | Sustainable Development Solutions Network |
| SE4All | Sustainable Energy for All |
| SEEA | System of Environmental-Economic Accounting |
| SEI | Stockholm Environment Institute |
| SIDS | Small Island Developing States |
| SR | "Synthesis report on best practices and lessons learned on the objective and themes of the United Nations Conference on Sustainable Development" |
| SRSP | Sarhad Rural Support Programme |
| *TEEB* | *The Economics of Ecosystems and Biodiversity* |
| TERI | The Energy and Resources Institute |
| TWN | Third World Network |
| UCLG | United Cities and Local Governments |
| UN | United Nations |

| | |
|---|---|
| UNCED | United Nations Conference on Environment and Development |
| UNCLOS | United Nations Convention on the Law of the Sea |
| UNCSD | United Nations Conference on Sustainable Development (Rio+20) |
| UN DESA | United Nations Department of Economic and Social Affairs |
| UNDP | United Nations Development Programme |
| UNECA | United Nations Economic Commission for Africa |
| UNECE | United Nations Economic Commission for Europe |
| UNEP | United Nations Environment Programme |
| UNESCO | United Nations Educational, Scientific and Cultural Organization |
| UNFCCC | United Nations Framework Convention on Climate Change |
| UNGA | United Nations General Assembly |
| UNGASS | United Nations General Assembly Special Session [to Review and Appraise the Implementation of Agenda 21] |
| UN-HABITAT | United Nations Human Settlements Programme |
| UNIDO | United Nations Industrial Development Organization |
| UNISDR | United Nations International Strategy for Disaster Reduction |
| UNSGAB | United Nations Secretary-General's Advisory Board on Water and Sanitation |
| UNU | United Nations University |
| UNV | United Nations Volunteers |
| WBCSD | World Business Council for Sustainable Development |
| WEO | World Environmental Organization |
| WMO | World Meteorological Organization |
| WRI | World Resources Institute |
| WSSD | World Summit on Sustainable Development |
| WTO | World Trade Organization |

# Introduction

The objective of this book is to give people an understanding of how the United Nations Conference on Sustainable Development (UNCSD), otherwise known as Rio+20, was conceived and developed, during one of the most difficult financial crises the world has seen.

As the book is written by three authors who took part in the process from the start – a government representative, a high-level UN official and a stakeholder leader – it gives a unique insight into a UN conference process and the internal and external dynamics that led to the final outcome. When Rio+20 had finished, all three of us felt that there was much that could be learned for future multilateral processes from what we and others had experienced. Looking around, there are few books written by people who have been part of intergovernmental processes. We hope that what we have written will give our readers a better understanding of the complexities of multilateralism in the first decades of the twenty-first century.

The book has been conceived in five chapters. Chapter 1, "The rebirth of sustainable development," reviews how Brazil and key other players slowly built a narrative for the rebuilding of sustainable development. It looks at the landscape in which Rio+20 was conceived – in particular, the financial crisis and how it played out in the United Nations and other fora. It looks at how the old order is changing, particularly the emergence of the BRICS countries (Brazil, Russia, India, China and South Africa), but also at the role of the ALBA (Alianza Bolivariana para los Pueblos de Nuestra América) group and other key emerging economies that during the past decade have often taken the driving seat in multilateral diplomacy. It picks up much of the story from the end of the book *Only One Earth* by Felix Dodds and Michael Strauss with Maurice Strong.

Chapter 2, "Chronicling the key impacts on Rio+20," explores the period between when it was agreed to hold the conference and the opening of the final preparatory meeting in Rio in June 2012. It explores both the formal and the informal roles that stakeholders played in the development of the zero draft and their contribution. It shows that, against the odds, and contrary to much-repeated media positions, Rio+20 developed a set of building blocks that may enable sustainable development to become the main conceptual framework for the twenty-first century.

Chapter 3, "How did it all come to happen?" explains the final two weeks in Rio: what the key divergences were when the delegates arrived in Rio, the conflicts within the Group of 77 and the role that the host country, Brazil, played in shepherding the process towards a consensus. It goes on to look at the non-formal outcomes from Rio+20, identifying a number that will continue to have an impact well into the future.

Chapter 4, "Multiplying commitments," looks at the multitude of other commitments and activities that happened in Rio around the formal negotiations. It picks out a number that we believe will continue to play a significant role in the future implementation of Rio+20 and help us in our quest for a more sustainable planet.

Chapter 5, "From Rio+20 to 2015 and the new development agenda," assesses how the outcomes from Rio+20 have been implemented and where they might have an impact on the development of new development goals to replace the Millennium Development Goals (MDGs) in 2015. It reviews the present changes in global governance and some that still need to be developed. Finally, it suggests what needs to be in place on January 1, 2016 if we are to show we have moved significantly in the right direction.

We have a number of people to thank for this book's development, in particular help on the editing from Dr Chantal Line Carpentier, Maruxa Cardama, Dan Coviello, Eela Dubey, Maria Figueroa Kupcuand, Chelsea Phipps and José Eduardo de la Torre Bárcena. We would also like to thank the following for their comments on particular parts of the text: Ambassador Patricia Espinosa Cantellano (Post-2015 Development Agenda), Sasha Alexander (desertification, land degradation and drought), Zachary Bleicher (food security, agriculture and rural development), Craig Boljkovac (chemicals), Maruxa Cardama (human settlements), Lee Cando (Global Sustainability Report), Jack Cornforth, Lalanath De Silva (Principle 10), Carole Excell (Principle 10), Owen Gaffney (science), Alexandra Garita (health and women), Gisbert Glaser (science), Harold Goodwin (tourism), Olimar Maisonet-Guzmán (water and sanitation), Andrew Higham (climate change), Jeannet Lingan (corporate social responsibility), Georgios Kostakos and Janos Pasztor (Global Sustainability Report), Dr Monika G. MacDevette (biodiversity), Emmanuel Print (sustainable consumption and production), Tracy Raczek (women), Anabella Rosemberg (employment, decent work for all and social protection), Dr Hesphina Rukato (Africa), Susanne Salz (human settlements), Kirsty Schneeberger (green economy), Mita Sen (forests), Richard Sherman (energy), Matt Simons (employment, decent work for all and social protection), Jan Gustav Strandenaes (governance), Vanessa Timmer (sustainable consumption and production), David Trouba (water and sanitation), Farooq Ullah and Zonny Woods (women), Alex Evans and David Steven (Sustainable Development Goals and G20), Nritya Subramaniam (global economic governance), René Orellana (rights of Mother Earth), Natalia Casas Mendoza and, at Earthscan, Tim Hardwick and Ashley Wright.

In this book you will find the names of many who have contributed in multiple and varying ways to a definition of, and the process of, sustainable development. We dedicate this book to those people whose names we know, but also to those whose names are unknown to us, who, by their effort and example, have contributed to the global understanding and advancement of a sustainable development agenda which embraces our shared planet and all its people with dignity, equity, justice and sufficiency. May their dream and our own become a reality.

We believe that the world is in a transitional phase with huge challenges ahead, many caused by the lack of political will to implement previous agreements. The book gives some ideas concerning a positive road map to the future. Through these challenging times the world needs a strong, robust and nimble United Nations, which we continue to believe has the credibility and capacity to be a force for global good and sustainability.

Many of the problems of energy security, food security, water security or economic security that we face will require multilateral solutions. The United Nations is the only legitimate body that represents *all* governments and is now, increasingly, the space for *all* stakeholders to bring their knowledge and expertise. Together we can solve these challenges for all people in a just and fair and equitable way. This book takes us on the journey to and from Rio+20 but it is up to all of us, working together, to make the leap into a more sustainable future. As Einstein said, "Learn from yesterday, live for today, hope for tomorrow. The important thing is not to stop questioning."

Chapter 1

# The rebirth of sustainable development

When Brazil started discussing the possibility of holding a conference on sustainable development in the autumn of 2007, it was in nobody's mind that such a conference could be a launching pad for a new global development partnership, one in which sustainable development is the basis for eradicating poverty, upholding human development and transforming economies. Such was the transformative role that Rio+20 played. Rio+20 represented a culminating point of decades of successes and failures of efforts aimed at defining a developmental pathway that reinforces the inescapable symbiosis between environment, human wellness and better standards of living.

The legacy of Rio+20 consisted of the agreement to develop integrated goals; renew environmental and developmental institutions; develop green and inclusive economic policy frameworks; and set principles and instruments for sustainable consumption and production, as well as frameworks for measuring development beyond the traditional gross domestic product (GDP). The rebirth of sustainable development draws its essence from two complementary approaches to development: a planet-centred approach and a people-centred approach. Both approaches find their common source in the principles agreed in the Charter of the United Nations. They have evolved distinctively during the past decades, through successive international conferences in the social, economic and environmental fields that have given birth to international norms, practices, instruments and institutions, often on parallel paths.

The planet-centred approach has had a number of key milestones in the past four decades in the area of environment and development. The United Nations Conference on the Human Environment, hosted in Stockholm in 1972, popularly called the Stockholm Conference, is regarded as the first multilateral conference on the environment. Stockholm gave impetus to national and international environmental protection, it ushered in the advent of Ministries of the Environment and launched activities to develop an intellectual and conceptual framework through which to view environmental challenges. It was also responsible for the creation of the United Nations Environment Programme (UNEP), gave rise to a body of international environmental institutions and to the creation of international environmental law, characterized by soft law[1] and hard law.[2]

Stockholm became the precursor to the international agenda of sustainable development, which was first promoted at its ten-year review held in 1982 in Nairobi, the seat of UNEP. This Special Session of UNEP's Governing Council promoted the concept of a UN Commission on Environment and Development. The then UN Secretary-General, Javier Pérez de Cuéllar, took up this idea and asked the Prime Minister of Norway, Gro Harlem Brundtland, to establish a World Commission on Environment and Development. The initiative was endorsed in December 1984 by the United Nations General Assembly (UNGA). This World Commission became known as the Brundtland Commission and published its report in 1987. The Commission's definition of sustainable development was in itself a paradigmatic revolution synthesizing development and environmental thinking. Sustainable development was defined as a balanced integration of social development, economic development and environmental protection that meets the needs of the present generation without compromising the ability of future generations to meet their own needs (United Nations, 1987).

Another key result from the Commission's report was the recommendation to hold a new conference on environment and development. This conference was agreed upon in December 1989 by the UNGA, which decided to convene a United Nations Conference on Environment and Development (UNCED) in 1992 – renamed by Maurice Strong as the Earth Summit – following a resolution drafted by two young development specialists from the South: Ahmed Djoghlaf from Algeria, who later became Executive Secretary of the United Nations Convention on Biological Diversity; and John W. Ashe from Antigua and Barbuda, who through the years rose to became his country's ambassador to the United Nations and then to occupy the highest elected office of the United Nations as president of the UNGA (2013–2014).

The Earth Summit has been the most significant event for advancing environment and sustainable development law and for creating a body of thought and literature on sustainability as a tool for development and well-being. Its outcomes have continued to frame our vision concerning sustainable development:

- Agenda 21;
- the Rio Principles;
- the United Nations Framework Convention on Climate Change;
- the United Nations Convention on Biological Diversity;
- the United Nations Convention to Combat Desertification in those countries experiencing serious drought and/or desertification, particularly in Africa;
- the Forest Principles;
- the United Nations Commission on Sustainable Development (CSD).

The people-centred development agenda embodies the foundations laid out in the 1948 Universal Declaration of Human Rights; the 1966 International

Covenant on Economic, Social and Cultural Rights – which recognizes the right to development – and the subsequent conventions and instruments that protect human dignity and freedom, equality, and equity of women, children, migrants, persons with disabilities, indigenous peoples and other groups. Some of the most significant conventions in this regard are the following:

- the International Convention on the Elimination of All Forms of Racial Discrimination (March 1966);
- the International Covenant on Civil and Political Rights (December 1966);
- the Convention on the Rights of the Child (November 1989);
- the International Convention on the Protection of the Rights of All Migrant Workers and Members of Their Families (November 1989);
- the Convention on the Rights of Persons with Disabilities (December 2006).

As Maurice Strong said at the end of the 1992 Earth Summit:

> The carrying capacity of our Earth can only sustain present and future generations if it is matched by the carrying capacity of its people and its leaders. We must bring our species under control, for our own survival, for that of all life on our precious planet.
>
> (Strong, 1992)

As a complement to the international human rights conventions, during the 1970s, 1980s and 1990s the United Nations also advanced the people-centred development agenda through major international conferences similar to those held in the fields of environment and sustainable development.

The improvement of gender equality and the empowerment of women gradually evolved through international conferences such as the first (Mexico, 1975), second (Copenhagen, 1980), third (Nairobi, 1985) and fourth (Beijing, 1995) World Conference on Women. These culminated with the adoption of the Beijing Action Platform, its follow-up and processes by the United Nations Commission on the Status of Women (CSW) and by new implementing institutions such as UN WOMEN, established in 2010 by the UNGA.

The apex of the conferences in the social field was attained in 1995 when the United Nations held the World Summit for Social Development in Copenhagen as a means for creating a dialogue space for confronting, in a holistic manner, the major social challenges of the decade. Some of these were in reaction to the changes in the world economy brought by globalization, the consequences of the structural adjustment policies in developing countries forced by the "Washington Consensus" (Serra and Stiglitz, 2008), the changing nature of the welfare state model, and the growing phenomenon of social disintegration in rapidly urbanizing metropolitan areas. The Copenhagen Declaration adopted at the end of the summit represented advancement towards an inclusive people-centred development agenda and introduced the idea of "social development goals" through the following commitments:

- promoting full employment as a basic priority of economic and social policy;
- promoting social integration and the promotion and protection of all human rights;
- achieving equality between women and men;
- promoting and attaining the goals of universal and equitable access to quality education and access for all to primary health care;
- accelerating the economic, social and human development of Africa and the least developed countries;
- ensuring that structural adjustment programmes include social development goals.

The culminating emblems of the people-centred agenda are without doubt the 2000 Millennium Declaration, the Millennium Development Goals (MDGs) and the Human Development Index (HDI). The Millennium Declaration is the ultimate global agenda for development and poverty eradication. It is based on six basic human values: freedom, equality, solidarity, tolerance, respect for nature and shared responsibility. Its poverty eradication chapter contains the pledge by world leaders to time-bound MDGs. The Millennium Declaration and the MDGs are a milestone in the promotion of a holistic approach towards human development, grounded in the historic opportunity offered by globalization and rapid technological change. Their purpose was to concentrate global attention and resources aimed at sparing

> no effort to free our fellow men, women and children from the abject and dehumanizing conditions of extreme poverty, to which more than a billion of them are currently subjected [and thus] making the right to development a reality for everyone and to freeing the entire human race from want.
>
> (United Nations, 2000)

The Millennium Declaration also contained a call to address in the 2001 Monterrey Conference on Financing for Development the main "obstacles developing countries face in mobilizing the resources needed to finance their sustained development" and the special needs of the least developed countries (LDCs) and other developing countries in special situations. It further charted an ambitious programme of institutional reform involving the UNGA, the Economic and Social Council and the cooperation and coherence between the United Nations, its agencies, the World Bank, the International Monetary Fund and the World Trade Organization, intending to do everything possible to ensure that the United Nations became

> a more effective instrument for pursuing the fight for development for all the peoples of the world, the fight against poverty, ignorance and disease; the fight against injustice; the fight against violence, terror and crime; and the fight against the degradation and destruction of our common home.

Finally, the Millennium Declaration also agreed to open greater opportunities to the private sector, non-governmental organizations and other stakeholders in general to contribute to the realization of the UN goals.

The Millennium Declaration, however, did not propose sustainable development as the overarching paradigm for achieving development and poverty eradication. The Declaration only refers to sustainable development in connection with "the respect for nature" and "the protection of our common environment." This dichotomy was an early indication of the difficulties faced at the start of the new millennium by the sustainable development paradigm and its association with an "environmental agenda."

By 2002 it was clear that the planet-centred development agenda faced real difficulties. Geopolitical factors, such as the aftermath of 9/11, would continue to detract attention from implementing the results of the World Summit on Sustainable Development held during that year in Johannesburg. Most importantly, the world's substantive commitment to development was elsewhere. The international community gathered in Johannesburg to look back on three decades of implementation since Stockholm and one decade since Rio, and to agree on a new and ambitious plan of action; the few advances made were quickly eroded. As early as 2006, President Mbeki of South Africa was already calling the Johannesburg Plan of Implementation (JPoI) a "forgotten piece of paper." In 2007, for the first time in its short history, the CSD failed to agree a policy outcome. In 2011 it failed again for a second and last time, since the CSD's life would come to an end after its 2011 meeting as a result of decisions made at the Rio+20 Conference.

The financial and economic crisis that started in 2008 also dealt a hard blow to the United Nations and the fulfilment of the pledges made at the dawn of the new millennium under both the people-centred and the planet-centred approaches. Its devastating effects on the prospects for global economic growth for 2009 and 2010 also acutely exposed the faltering coordination and systemic surveillance by the Bretton Woods institutions and the United Nations. These circumstances called into question the overall relevance of the United Nations in the global setting and its ability to deliver on the pledges made to achieve poverty eradication and sustainable development.

Despite the various reforms undertaken particularly from 2005 onwards as a result of a major reformist programme agreed in the first five-year review of the MDGs, known as the World Summit Outcome, by the end of the decade the United Nations was losing its relevance to act as a global coordinator capable of responding to the major evolving social, economic and environmental situations, and attracting and bringing to the same table all actors involved in the solution to those problems. The United Nations was at pains to deliver to the world a coherent message with regard to its own global development priorities. This situation was partially due to a growing internal tendency to work in silos. It was also the result of the inability of UN member states to come to terms with the ambiguity previously pointed out between the human-centred "development and poverty eradication" agenda and the planet-centred agenda, which often led to the erroneous association of "sustainable development" as simply an "environmentalist agenda."

## President Lula da Silva's and Brazil's case for a Rio+20

Taking up the baton for sustainable development in light of President Mbeki's comments, on September 25, 2007 President Luiz Inácio Lula da Silva, addressing the UNGA at the opening of its annual leaders' week, made the case for a new conference on sustainable development (UN News Centre, 2007).

President Lula indicated the necessity for a summit that would "address the persistent problem of global inequality," asserting that "social equality is our best weapon against the planet's degradation." He stressed the need to reorder international priorities around an agenda that supported and favoured "social justice." Recalling Rio's efforts to preserve and protect "our common heritage," President Lula stated that this heritage could only be "salvaged through a new and more balanced distribution of wealth." He called for greater emphasis on renewable energy and biofuels, for international trade relations based on balanced and fair rules, and for development approaches which would result in the termination of farm subsidies that make the rich richer and the poor poorer, agricultural protectionism and the perpetuation of "dependency and underdevelopment." President Lula stated, "[I]t is unacceptable that the cost of the irresponsibility of a privileged few should be shouldered by the dispossessed of the earth."

On November 4, 2008, over a year after the address by President Lula, the government of Antigua and Barbuda tabled a draft resolution on behalf of the group of developing countries (the Group of 77) for the 63rd session of the United Nations General Assembly, calling for a World Summit in 2012. The original proposal expressed deep concern that

> although some progress has been achieved since the landmark Conference of 1972, the Earth Summit in 1992 and the World Summit on Sustainable Development in 2002, there is a persistent implementation gap, and many commitments by the international community have not been fully met.
> (United Nations, 2008b)

On November 11, 2008, in the framework of the UNGA negotiations, Brazil disseminated a Concept Note indicating how its delegates saw the conference, its potential, purpose and outputs. It was Brazil's view that Rio+20 was intended to "create the political momentum for concentrating disparate actions, channeling efforts and taking broad-reaching decisions on the agenda of sustainable development" (Brazil, 2008). It was to be "a stimulus for action." It was envisaged that the conference would focus on (1) the Rio Declaration, (2) the international environmental conventions, and (3) Agenda 21 – specifically, how these were being addressed and implemented, and what remained outstanding in the international agenda. Its purpose was also to "improve coordination and cooperation," review international institutions in a manner that "might conclude the debate on international environmental governance," and strengthen the role of stakeholders in the implementation of the global agenda on sustainable development.

It was anticipated that Rio+20 would re-evaluate, support and strengthen the role of the CSD, "providing political guidance; promoting dialogue and partnerships and fostering coherence between implementation and actions." Brazil insisted that the debate should not be primarily environmental but multidimensional and cross-disciplinary, reflecting the three intersecting pillars of sustainable development. While it envisaged the strengthening of international organizations, Brazil recognized that "the needs of all countries should be respected," that national and regional institutions also required similar strengthening, commensurate with what would take place at the global level. Brazil made it clear, however, that, unlike in the 1992 situation, "a new conceptual framework was not intended as the main outcome."

In its 2008 Concept Note, Brazil raised its concerns on the marginalization of the Commission on Sustainable Development as the locus for setting the agenda for sustainable development and supporting national efforts in rationalizing sustainability. The initial view of Brazil, which was shared by the vast majority of developing countries, was supportive of the CSD. These member states thought that the CSD should be "strengthened and recognized as an important UN body." The intent was to have Rio+20 "build on and complement work on the CSD" and to have the Conference "provide a mid-term appraisal of the commission's work cycle." Brazil also saw the 2012 conference as "an important stepping stone to the MDG review in 2015," although it did not specify its perception of the relationship between the two milestones. In this context, Brazil articulated that the Rio+20 Conference was not intended to reopen settled documents, agreements and conventions, and would not seek to take over current negotiations taking place elsewhere, such as the climate change negotiations. Rio+20 would instead allow "a strong and urgent political signal from the United Nations General Assembly that sustainable development continues to merit the highest political attention." The key deliverables from the perspective of the host country were to be:

- to review high-level political commitments to sustainable development;
- to provide an opportunity to seek consensus on areas of sustainable development that have evolved considerably in the past decade, such as international governance and the production and use of renewable energy;
- to address gaps in the implementation of Agenda 21 and Multilateral Environmental Agreements;
- to strengthen the role of governments, Major Groups and other stakeholders such as the private sector through partnerships of the sustainable development agenda (Brazil, 2008).

It is reasonable and fair to say that there was no initial support from developed countries for the idea of a summit on sustainable development. The 2008 proposal was not immediately retained. Instead, the UNGA opted for taking a step back and invited "Member States to express their views on the possibility of convening a high-level event on sustainable development" (IISD Reporting Services, 2008).

This door opened by the UNGA, allowing further reflection on how to review the status of implementation of the sustainable development agenda, was used by stakeholders, governments and some intergovernmental bodies meeting on November 11–13, 2008 in Donostia (San Sebastián), Spain. That setting allowed the initial thinking behind a conference on sustainable development to be reformulated and additional priorities for the conference to be advanced. Some of the new ideas that were developed by the end of 2008 included:

- the green economy;
- new and emerging issues;
- a review of international and member states' commitments;
- institutional reform (Stakeholder Forum, 2008).

It was not until Christmas Eve – December 24, 2009 – that the General Assembly finally adopted (without a vote) its Resolution 64/236, accepting Brazil's offer to host a conference to follow up on the 1992 Earth Summit (United Nations, 2009b). This final outcome had required two years of sustained efforts by Brazil and many other stakeholders. In the intervening period between President Lula's speech in 2007, the issuance of the concept note in 2008 and the UNGA Resolution of 2009, Brazil had been engaging with UN member states to secure agreement on the hosting of a conference and acted as the champion of the cause. Once the decision to have the conference was taken, Brazil adjusted its role, and to a certain extent yielded, to the pace set up by the preparatory committee and its governing structures. The process laid out by Resolution 64/236 created a coordinating body for the Rio+20 Conference, which was presided over jointly by Antigua and Barbuda and the Republic of Korea, respectively represented by Ambassador John W. Ashe and Ambassador Kim Sook (who was preceded by Ambassador Park). Both of them – along with the eight other members of a Bureau drawn from each of the five UN regions, plus Brazil, which participated as an ex officio member – were charged with the responsibility of guiding the process, promoting a multi-stakeholder dialogue and starting the logistical and other arrangements for hosting the conference.

Over the two and a half years between the UNGA Resolution and the actual conference, the role of Brazil as proponents and hosts of Rio+20 would continue to evolve and would reach its apex when the life of the Bureau came to an end, at which time Brazil took over the negotiations, secured a consensus text and brought to a successful close the largest multilateral conference in the history of the United Nations.

*Figure 1.1* Ambassadors John W. Ashe and Kim Sook during the second round of informal negotiations, April 23, 2012–May 4, 2012 (courtesy of IISD/*Earth Negotiations Bulletin* (www.iisd.ca)).

## The 2008 crisis as a catalyst for breaking the silos between development and sustainable development

The Rio+20 Conference preparations were happening in the aftermath of possibly the greatest financial crisis the world had seen since the 1920s. The collapse of Lehman Brothers – one of the major global financial investors – in mid-September 2008 quickly spread throughout the different layers of the international financial system, shaking the institutional foundations of global economic governance. The United Nations was not the only organization caught unprepared to deal with what appeared to be the deepest systemic recession since the 1920s. The International Monetary Fund (IMF) was also unprepared. The flagship 2009 report of the IMF opened with a puzzled "How did things get so bad, so fast?" (IMF, 2009).

The economic shock was unprecedented. During the fourth quarter of 2008, advanced economies experienced a 7½ per cent decline in real GDP. Emerging developing economies proved more resilient to the impact of the crisis, but, even so, they were negatively exposed, owing to the shrinking of their trade perspectives and a lack of financial influxes. The LDCs felt the burden of the crisis severely. Small Island Developing States (SIDS), graduating from concessionary financing and shrinking official development assistance (ODA), already grappling with high debt to GDP ratios and adverse climate change impacts, and vulnerable to external shocks, both environmental and economic, were particularly hard hit. Several factors contributed to the unequal impact among developing countries. The volatility in global commodities markets triggered by the crisis, coupled with the high levels of indebtedness and financial exposure, jeopardized the LDCs' prospects for recovery. The fragility and lack of resilience within LDCs was also set to undermine much of the developmental gains achieved through the MDGs.

The impact of the crisis and the uncertainty it triggered concerning the future prospects for development was not confined to the macroeconomic and financial spheres. It also shook the political and institutional dynamics within multilateral institutions, starting with the United Nations and the IMF. Ever since the news of the shock in the New York Stock Exchange had spread, it had been widely felt that globalization and interdependence had played a critical role in the widespread nature of the crisis.

The crisis exposed an emerging divide among decision-makers on the deficient surveillance and monitoring by international institutions, particularly in those advanced economies where the collapse had taken place. The crisis also gave rise to another debate about which institutions were the best placed to coordinate and lead in restoring international confidence and addressing the structural factors behind the crisis. One vision, dominant among finance ministries and central bank governors, coalesced around the embracing of a flexible and decentralized global approach by putting effectiveness at the forefront of the response. This approach had initially led, on September 25, 1999, to the creation of the Group of Twenty (G20) forum of finance ministers and central bank governors as an informal global policy coordination mechanism. It alleged that the underlying causes of the crisis were the insufficiently coordinated macroeconomic policies, the lack of structural reforms and weak domestic regulatory oversight. Others, however, had a completely different perspective. The alternative view had a strong foothold among representatives of ministries of foreign affairs, particularly those in a wide variety of middle-income countries, who favoured a more political approach towards the crisis and its root causes. According to this point of view, the 2008 financial collapse was just part of broader structural and systemic deficiencies that required attention beyond the scope of the financial regulation and oversight. This vision highlighted at the global level the need to strengthen legitimate decision-making platforms and develop more democratic processes of collective consultation. It also emphasized a need to proactively involve developing countries in the identification of solutions.

Although held within only a few weeks of each other, the summary of the UN Interactive Panel of the Global Financial Crisis – convened by the president of the UNGA, at that time Miguel d'Escoto Brockmann from Nicaragua (United Nations, 2008) – and the declaration of the G20 Summit on Financial Markets and the World Economy held in Washington, DC, in November 2008 (G20, 2008) provide an early illustration of this divide. The latter made reference to the United Nations only in connection to the MDGs and the development principles agreed in the 2002 Conference on Financing for Development. The former made an "urgent call" for the UN – labelled then as "G192" – to play a central role in the search for solutions and "restore legitimacy to the international financial system." Despite the differences of approach between the first meeting at the level of leaders of the G20 and the initial discussions spearheaded by the Panel of Experts convened by the UNGA under President d'Escoto Brockmann, there were also common points. The first one was the common concern for correcting the deficits in the governance of the global financial architecture, particularly within the IMF. Although the two documents advocated different mechanisms, both visions also called for significantly improved coordination between the United Nations and the international financial institutions. The third common point was the urgency of maintaining the efforts being made to implement the MDGs.

### *Breaking the silos between development and sustainable development*

The 2008 crisis also acted as a catalyst for breaking within the UN the silos between the follow-up of commitments associated to development and those associated to sustainable development. This gradual normative and institutional convergence was also enhanced by the increasing political relevance of the climate change negotiations and by the increasing realization by policy makers and negotiators acting on their behalf that climate change cuts across the social, economic and environmental priorities of any national developmental model. In addition to these, other factors also influenced this convergence. These include the evolution of traditional development cooperation, the increased availability of scientific information supporting nexus models and integrative development policies. All of these catalyzed a rethinking within the UN of its norm setting processes relating to development and its social, economic and environmental dimensions. Each one of the norm-setting and review conferences that will be reviewed in this section contributed to build a bridge between development and sustainable development and between the planet-centered and the people-centered agenda culminating in Rio+20 and ultimately in the foreseen development summit in September 2015 where a framework and set sustainable development goals will be agreed.

The institutional background against which the United Nations decided to convene between 2008 and 2012 four major international conferences aiming – from different angles – at restoring the relevance of the United Nations as an international organization responsible for enhancing systemic policy coherence was complex and deeply fragmented. A first testing experience of this growing divide was seen in the dynamics behind the negotiations of the Doha Declaration during the United Nations Review Conference on Financing for Development, held in Qatar from November 29 to December 2, 2008. This conference was followed by the 2009 Conference on the World Financial and Economic Crisis and Its Impact on Development and by the 2010 High-Level Plenary Meeting of the United Nations General Assembly on the MDGs. These three major conferences can be seen as attempts from different perspectives to advance system-wide issues left unresolved by the Millennium Declaration and the 2005 World Summit review.

Fulfilling the potential impact that the United Nations can have in global development has been a constant concern of the major reform initiatives undertaken in the United Nations, particularly since the UN Secretary-General Kofi Annan's 2005 manifesto entitled *In Larger Freedom: Towards Development, Security and Human Rights for All* (United Nations, 2005a). However, behind most calls for reforming the role of the United Nations in the area of economic governance there has always been a tension between a narrow and a broad definition of the organization's scope, depending on the interpretation given to the coordination powers provided to the Economic and Social Council (ECOSOC) by article 63 of the Charter of the United Nations with regard to the specialized agencies, in particular the Bretton Woods Institutions.

Annan's reformist programme provided the conceptual backbone for the 2005 World Summit that was convened to carry out a first five-year assessment of the commitments undertaken in the Millennium Declaration. The World Summit Outcome – the official title of the declaration adopted at the end of the summit by the world leaders – stands as one of the most significant reform-oriented frameworks set out by the UNGA for each of the three pillars of the organization. In the human rights pillar it led to the establishment of the Human Rights Council. In the peace and collective security pillar the Outcome complemented the scope of the peacekeeping activities of the United Nations with an approach focused on peacebuilding. The Peacebuilding Commission was therefore established as the dedicated mechanism to "bring together all relevant actors to marshal resources" for a "coordinated, coherent and integrated approach to post-conflict peacebuilding and reconciliation." In the development pillar of the United Nations, the Outcome provided for a first reform of ECOSOC's structure and functions in the light of its increasing relevance in the following up of the MDGs. Also in the development pillar, the Outcome reconfigured the role of ECOSOC and the UN Secretary-General (1) in the coordination of the UN operational activities for development through the "Delivering as One" pilot initiative; (2) in the coordination of humanitarian assistance through the reconfiguration of the operation of the Central Emergency Response Fund, and the establishment of a humanitarian segment within ECOSOC; and finally (3) in the integration of the environmental activities of the United Nations in the broader sustainable development framework, thus giving way to the debate on international environmental governance (IEG).

The Outcome also expanded some of the commitments set out in MDG 8 dedicated to a Global Partnership for Development. It improved national ownership by requesting all countries to define by 2006 "national development strategies" to achieve the MDGs. The Outcome also spelled out further commitments in the different chapters of the financing for development agenda, particularly in official development assistance (ODA), domestic resources mobilization, investment, debt, trade and commodities. Finally, the Outcome advanced a definition of what role world leaders expected the United Nations to play in global economic governance. In its chapter entitled "Systemic issues and global economic decision-making," the agreement opted for a development-oriented definition which recognized that the United Nations plays

> a fundamental role in the promotion of international cooperation for development and the coherence, coordination and implementation of development goals and actions agreed upon by the international community [and to this end committed] to strengthen coordination within the United Nations system in close cooperation with all other multilateral financial, trade and development institutions in order to support sustained economic growth, poverty eradication and sustainable development.
>
> (United Nations, 2005)

The World Summit Outcome opened a door to another wave of UN reform which coincided with the end of the mandate of Secretary-General Kofi Annan. This last reformist effort was driven by a High-Level Panel on United Nations System-Wide Coherence. Three serving prime ministers co-presided over this effort: Prime Minister Shaukat Aziz of Pakistan, Prime Minister Luísa Dias Diogo of Mozambique and Prime Minister Jens Stoltenberg of Norway. It also comprised a secretariat headed by Adnan Amin, one of the most effective UN directors, whose work was recognized by later going on to lead the International Renewable Energy Agency (IRENA). Their report to the outgoing Secretary-General advocated providing the United Nations with additional tools to better perform a systemic surveillance, monitoring and agenda-setting role building on the assets of the General Assembly "as the highest intergovernmental body" and ECOSOC as the "main body for coordination, policy review and policy dialogue." The System-Wide Coherence Report advocated enhancing the potential of the Council by establishing a Global Leaders Forum – an initiative which was not implemented at the time but which stimulated the conceptual architecture of the High-Level Political Forum in the run-up to Rio+20. The Leaders Forum, as advocated by the Panel, should "comprise the leaders of half its members, rotating on the basis of equitable geographic representation, with the participation of the executive heads of the major international economic and financial institutions." It suggested that its meetings be preceded by a preparatory meeting of ministers for foreign affairs and economic, social and related ministries. The role of such a Global Leaders Forum would be to provide leadership and guidance to the international community on development- and global public goods-related issues, and to develop a long-term strategic policy framework to secure consistency in the policy goals of the major international organizations (United Nations, 2006).

In the wake of the 2008 crisis, the coordination deficiencies that had become clear since 2006 put the United Nations in a difficult position to deliver to the world a coherent message on recovery, coordination and development. This situation, as we have already indicated, only exacerbated the growing tendency to consider development issues in silos. This came as a result of the inability by UN member states to come to terms with the internal inconsistencies between the "development and poverty eradication," "sustainable development" and "global economic governance" agendas.

The lack of global political clout and relevance by the United Nations, which has been provided since 2008 by the G20, was not the only challenge that the crisis exposed. The gaps between the global, the regional and the national levels were also made visible. For some of the real-world challenges, in the midst of various crises the United Nations seemed only capable of providing global and "standardized" objectives which were perceived as too abstract and lacking in concrete indications on how to achieve them.

The 2008 crisis also exposed a need for areas of flexibility within the intergovernmental landscape, where systems can evolve and allow for differentiated implementation in accordance with particular national situations, while remaining overall compatible with global frameworks and principles.

There has been a shifting focus from global objective-setting coupled with a gradual movement towards an increased reliance on creating incentives to address poverty eradication and sustainable development. This has relaunched a creative reflection on which institutions were best prepared to bridge gaps in implementation, particularly through better engagement with stakeholders to enable countries to translate global development objectives into policies and actions. In some cases such gaps had already been filled by regional development banks and regional fora, where cooperation seemed easier and peer pressure stronger.

The substantive support functions of the United Nations to help member states integrate social, economic and environmental policies were questioned in the context of the economic crisis. The coordination between UN agencies and other international organizations was felt to be insufficient, despite the availability within the system of means to provide relevant information and increased collaboration within and among the economic, environmental and social pillars.

After 2008 there was a slowing down of efforts to achieve the MDGs and a shift in the cooperation patterns of new development initiatives by donor countries. The financial crisis also spelled out some of the major challenges for the United Nations in this field, including its inability to provide the leadership, agenda-setting and the integration needed to support increased implementation at the national and global levels; to provide relevant knowledge; to strengthen effective system-wide coherence and coordination; and, finally, to involve all relevant actors and stakeholders. Against this backdrop, the ramifications of the global economic crisis also contributed to creating within the United Nations a pervasive lack of trust among key political players and a growing realization of the artificial nature of the divide between the so-called development and sustainable development agendas. The contraction of resources allocated to international cooperation and the perception that countries may step back from previously agreed commitments exposed a growing mistrust between developed and developing countries. This situation undermined the expectations and results of most if not all multilateral processes in the area of development convened between 2008 and 2012. In January 2009 the then UK prime minister, Gordon Brown, expressed the stark choices when he said:

> We face a choice. We could allow this crisis to start a retreat from globalisation. As some want, we could close our markets – for capital, financial services, trade and for labour – and therefore reduce the risks of globalisation. But that would reduce global growth, deny us the benefits of global trade and confine millions to global poverty.
>
> Or we could view the threats and challenges we face today as the difficult birth-pangs of a new global order – and our task now as nothing less than making the transition through a new internationalism to the benefits of an expanding global society – not muddling through as pessimists but making the necessary adjustment to a better future and setting the new rules for this new global order.
>
> (Summers, 2009)

### *The emergence of the Group of Twenty (G20)*

Despite the United Nations' early attempts to articulate a response to the 2008 crisis, as developments unfolded it became clearer that the organization was on the brink of losing its relevance as the authority for global decision-making processes. The basic ability of the United Nations through its principal organs, such as the General Assembly and the Economic and Social Council, to seat all relevant actors around the same table to deal expeditiously with a problem was simply not perceived as a credible option for coordinating macroeconomic and financial issues. As the end of the first decade of the twenty-first century approached, this crisis of confidence also undermined the prospects of the United Nations to find solutions in the sustainable development agenda.

In 2008, before any decision had been made to convene Rio+20, we witnessed the emergence of the G20 as one of the central informal global policy coordination mechanisms. Despite its existence since 1995 as a coordination mechanism for finance ministers, the upgrading of the G20's meetings – particularly after its first meeting of leaders convened in November 2008 – was perceived as a threat to the systemic relevance of the United Nations. The decision by the US President, George W. Bush, to convene, on an extraordinary basis, the leaders of the G20 fulfilled a function that the United Nations had failed to achieve: to provide a space – albeit limited – for world leaders to replicate at the global level the arbitration function between social, economic, financial and environmental ministries. The G20 emerged as a logical expression of a fragmented institutional system. It gradually became a forum for nations looking to tackle issues of financial stability in which the eight countries of the G8 no longer represent a critical mass and where consensus-building among the 190-plus member states within the United Nations militates against greater agility. However, the international economic system is now too complex for centralized oversight and decision-making. It is not a system of one wheel with many cogs. It is many wheels with many cogs (Pérez Verdía-Canales, 2013)

After several years of gradual evolution, from 2009 onwards the G20 has become a complementary and supportive space to many of the priorities that are set in the context of the United Nations. This has been the case in the drive towards the green economy, climate change financing, food security, social protection floors, sustainable energy and fossil fuel subsidies reform, which have been incorporated as part of the past presidencies of the Group – in particular those of Republic of Korea (2010), France (2011), Mexico (2012) and Russia (2013) – and therefore helped the Rio+20 Conference address these issues.

### *The 2008 Doha Review Conference on Financing for Development*

The Doha Declaration and its immediate aftermath, the 2009 United Nations Conference on the World Financial and Economic Crisis and Its Impact on Development, played an important role in setting the political dynamics that would prevail throughout the Rio+20 preparatory process. They were also instrumental in breaking the silos between the people- and the planet-centred development agendas.

The Doha Review Conference was not conceived to provide a platform for confronting the challenges posed by the economic and financial crisis. At first glance, the Doha Declaration which resulted from the conference seems even to overlook the gravity of the economic and financial situation and its impact on sustainable development. Its first sections, drafted and agreed upon during the first three-quarters of 2008, are a detailed reflection of the progress and challenges in the implementation of the Monterrey Consensus and the financing for development agenda before the economic crisis. Despite an orderly process of negotiations for the Declaration, as the Doha Conference opened with the attendance of more than 160 delegations – including of close to 40 heads of state or government – several key unresolved issues remained unresolved and jeopardized the conclusion of the conference itself. The financing for development agenda covered six broad areas, which are reflected in the Doha Declaration (United Nations, 2008a) as follows:

1 the mobilization of domestic financial resources;
2 the mobilization of foreign direct investment and other private flows;
3 the use of international trade as an engine for development;
4 official international financial and technical cooperation;

5 external debt and the coherence of the international monetary system;
6 financial and trading systems.

However as the year-long preparatory process for the conference evolved, it became clear that the global landscape was shifting and the United Nations ran the risk of being marginalized. The presidents of the negotiations concerning the Declaration – Maged A. Abdelaziz, Egyptian ambassador, and Johan Løvald, Norwegian ambassador, as well as Oscar de Rojas, a distinguished former Venezuelan diplomat who was appointed director of the UN Financing for Development Office and executive secretary of the conference – understood that Doha risked being sidelined by the magnitude of a crisis which required transformational approaches. This call was reflected in the concluding chapter of the Doha Declaration, entitled "Other new challenges and emerging issues."

A first set of issues was the consensual acknowledgement of the impact of the financial crisis and of the global economic slowdown on the ability of developing countries to access the necessary financing for their development objectives. Such recognition was not achieved until the very last moment since it was intertwined with other highly conflicting issues. One of these was how to define the share of the collective burden of overcoming the crisis that middle-income countries should bear. Another highly contentious issue, in particular among middle-income countries, was a consensual agreement on greater accountability with regard to tax havens and on investment flows. Finally, Doha almost failed on the issue of identifying basic principles of aid effectiveness to South–South cooperation activities.

The Doha Declaration also allowed the addressing of a host of new challenges that had not been identified before the 2008 crisis struck. There was an absence of mechanisms to protect the developmental gains achieved through the implementation of MDGs. This was because the concept of social protection floors had not been sufficiently developed. The Doha Declaration reflected with greater clarity than the early G20 declarations the need to learn from the painful social setbacks suffered by the structural adjustment experiences of the 1990s. In doing so, it drew a critical conclusion which was going to be later fully developed in the 2010 review of the MDGs: that the international financial institutions, particularly the IMF, had lacked adequate mechanisms to facilitate the integration of social gains in its structural adjustment plans. This lesson eventually encouraged greater inter-organization collaboration by the international financial institutions with the International Labour Organization (ILO) at the technical level – so that structural economic policy formulation would in future be mutually influenced upstream by social analysis.

Doha also recognized that the framework of the Monterrey Consensus and the financing for development agenda had so far focused solely on socio-economic challenges and had failed to take into account environmental sustainability, or climate change mitigation and adaptation. In doing so, Doha placed particular emphasis on the additional financing implications for sustained action for climate change, which materialized with the establishment of the

Green Climate Fund in the 2010 Climate Change Conference in Cancún. It also opened a door through which to address, within the United Nations, the relationship between energy markets and sustainable development. In doing so, it identified the need to promote energy systems consistent with climate change goals and spelled out the critical triple relationship of access to basic energy services, increase of renewable forms of energy and promotion of energy efficiency and conservation in connection with the acceleration of the implementation of the MDGs.

A last element raised by the concluding section of the Doha Declaration reflected deep concern about how to remain engaged in the social, economic and environmental areas to reflect the realities and challenges of the twenty-first century; while at the core of one of the harshest crises was the insufficiency of the international structures.

The 2008 Doha Conference was followed by the 2009 Conference on the World Financial and Economic Crisis and Its Impact on Development and, in turn, by the special meeting on MDGs in 2010. In this sense, Rio+20 was the fourth in a line of UN conferences held after 2008 which, from different perspectives, have defined the current role of the United Nations in a changing institutional landscape.

### *The 2009 Conference on the World Financial and Economic Crisis and Its Impact on Development*

The decision to hold another conference– on the world financial and economic crisis and its impact on development – less than a year after the Doha Review Conference was not only a real achievement by the Group of 77 but also the culminating moment of the UNGA presidency of Miguel d'Escoto. His period was marked by a strong will to affirm, in difficult and uncertain times, the central position of the United Nations as the only institution capable of bringing together all views and perspectives.

Despite diverging opinions on the extent of its legacy, the 2009 summit allowed a growing convergence of many of the conceptual streams that advocated for long-term solutions to the crisis and for reinvigorating the role of the United Nations as the driving force behind the global development agenda. It also laid the ground for the current configuration of a variable geometry in which the United Nations, the World Bank and the IMF, among other actors, support global development efforts in a reinforcing manner. Finally, it sent the message that short-term stimulus packages being introduced at the time in many developed and middle-income countries had to work for all, including the most vulnerable economies, and had to take a longer-term perspective, thus opening an opportunity for "inclusive green new deals."

The outcome of the 2009 summit was successfully agreed early on, owing to the work of the ambassadors of the Netherlands and Saint Vincent and the Grenadines. A week before the conference started in New York, they submitted for the consideration of UN member states a clean draft that was based on the inputs received during several rounds of consultations. This text was then swiftly adopted. This allowed a clear and ambitious narrative to be retained. The stellar work of the facilitators was helped by the leadership of the Group of 77. The outcome text starts out by stating that the United Nations, "on the basis of its universal membership and legitimacy, is well positioned to participate in various reform processes aimed at improving and strengthening the effective functioning of the international financial system and architecture." It then spells out all the major development impacts and outlines more than ten actions for a "coordinated and global response to the crises" (United Nations, 2009). Among the responses it identified were:

- safeguarding economic, development and social gains and providing adequate support for developing countries to address the human and social impacts of the crisis;
- fostering an inclusive, green and sustainable recovery, and providing continued support for sustainable development efforts by developing countries;
- strengthening the role of the UN development system in responding to the economic crisis and its impact on development;
- reforming and strengthening the international financial and economic system and architecture, as appropriate, to adapt it to current challenges (ibid.).

The 2009 conference and its resulting political declaration was the first opportunity the United Nations had to highlight the linkages between proactive recovery policies, defined as "global green initiatives," and the promotion of sustainable development. It signalled that the crisis should not "delay the necessary global response to climate change and environmental degradation, taking into account the principle of common but differentiated responsibilities and respective capabilities" (United Nations, 2009). On the contrary, the crisis presented an opportunity to promote "green economy initiatives" that could be implemented in national stimulus packages. These initiatives were seen as contributing to "achieving sustainable development, sustainable long-term growth, promotion of full and productive employment and decent work for all and poverty eradication" (ibid.).

As part of the preparations for the conference, the president of the UNGA, Miguel d'Escoto, convened a Commission of Experts on reforms of the international monetary and financial system. The commission was composed of 17 experts headed by the Nobel laureate Joseph Stiglitz and in September 2009 it produced *Report of the Commission of Experts of the President of the United Nations General Assembly on Reforms of the International Monetary and Financial System*. A second initiative focused on integrated measurement of development was convened in February 2008 by the president of France and was named the Commission on the Measurement of Economic Performance and Social Progress, otherwise known as the Stiglitz–Sen–Fitoussi Commission. The Commission of Experts took over some of the system-wide policy-coherence issues confronted by the 2006 high-level panel, and presented its own independent contribution to the World Financial and Economic Crisis conference. The Stiglitz Commission report, like its 2006 predecessor, argued that the United Nations can in fact live up to the challenges of, first, identifying effective and integrated modalities for promoting development in its broadest sense and, second, thus achieving a better integration of the economic dimension into the social and environmental dimensions of sustainable development.

The Stiglitz Commission nevertheless took a more detached systemic view over the UN institutional landscape and refrained from advocating specific reforms within the UN machinery. First, it supported increasing the membership and voting rights in the IMF and the World Trade Organization (WTO). It also argued for creating innovative agenda-definition spaces where representation and democratic legitimacy could be combined with limited membership, effectively chosen through democratic mechanisms, and with equitable representativeness. The Stiglitz Report did in fact propose the creation of a new institution, the Global Economic Coordination Council, to be supported by an independent "International Panel of Experts" tasked with the assessment and monitoring of both short-term and long-term systemic risks in the global economy (Stiglitz, 2010).

The conference did not yield a transformed institutional framework. It did set up a short-lived open-ended working group under the General Assembly which produced a progress report on the following up of the issues contained in the Outcome of the Conference on the World Financial and Economic Crisis and Its Impact on Development. Its conclusions were instrumental in improving the governance and accountability of the Bretton Woods Institutions, in particular the IMF and the WTO. They also contributed to regularizing the consultations between the G20 and the broader UN membership before each of the G20 summits, as part of an enhanced mutual outreach policy. The conference also allowed a "variable geometry" configuration in the G20 decision-making process to allow non-G20 states to participate in the G20 process. Finally, it opened a debate on how the United Nations could be reformed to respond to the challenges posed by the integration of the three pillars of sustainable development.

### *The 2010 High-Level Plenary Meeting of the General Assembly on the MDGs*

The strain imposed by the crisis on the international cooperation budgets and the discussions held for the Doha Review and the world financial and economic conferences compelled the development community to initiate an early dialogue on the gaps and the lessons drawn from the implementation of the MDGs development framework. The setting for such assessment was provided by the 2010 review of the MDGs, which was held in conjunction with the leaders' week of the UNGA, thus resulting in a highly visible and politically charged discussion.

The key message that was captured in its political outcome, entitled "Keeping the Promise: United to Achieve the Millennium Development Goals," was the growing realization that the achievement of the MDGs was not only a matter of aid and resources. Success in the implementation of the MDGs, we became aware, required levelling of the playing fields for trade and for the financing and investment conditions in developing countries, and, most importantly, "enhancing at all levels the policy coherence for development" (United Nations, 2010).

The first decade of implementation of the MDG framework had helped decision-makers realize that success in any of the eight global goals required coherent, "mutually supportive and integrated policies across a wide range of economic, social and environmental issues for sustainable development" (United Nations, 2010). In order to achieve the MDGs, "countries had to formulate and implement policies consistent with the objectives of sustained, inclusive and equitable economic growth, poverty eradication and sustainable development." A critical advance by the 2010 review was the definition of sustainable development as the path for the pursuit of the goals of eradicating poverty and hunger. This simple acknowledgement signified a radical departure from the original vision of the 2000 Millennium Declaration, which was grounded in committing developing countries' governments to a particular type of growth and development path, one that is equitable and human rights-based but not necessarily sustainable.

The assessment of the MDGs developed after the 2010 review also allowed the singling out of key challenges that the framework had failed to address. Some of these gaps, which later on formed the conceptual framework of the Post-2015 Development Agenda, were identified by the UN Committee for Development Policy (CDP):

- There have been changes in worldwide demographic dynamics and a transformation of the age structure of populations.
- Economic growth has not automatically reduced inequalities at the global and national levels.
- Environmental pressures on the life-supporting functions of the planet have increased.
- Instabilities in global markets and the lack of global mechanisms to deal with simultaneous crisis are now blatant.
- Inadequacies of governance in responding to the voices of people demanding fair, participatory and democratic approaches have come to the surface.
- A fragmented, increasingly complex aid architecture is now in place (CDP, 2012).

## The old order changes and new players in the developing world emerge

The convergence around the paradigm of sustainable development embodied by Rio+20 and by the UN conferences previously reviewed was made possible by the gradual refocusing on environmental thresholds by the development community. This comprises development agencies, development non-governmental organizations and agencies, development ministries and developing countries. This refocusing owes everything to major changes in geopolitical dynamics. The economic and financial crisis of 2008 provided a golden chance for a growing number of developing countries to redouble their claims to a greater share of influence in international decision-making and coordination processes with a direct impact on their own national development perspectives. The increasing influence of emerging countries can be seen everywhere: in the evolution of the climate change negotiations, particularly after the conferences in Cancún, Durban and Doha; in the reform and democratization of the governance and the quota system of the IMF and the World Bank; in the relentless stalemate of international trade negotiations within the Doha Development Round; and, last but not least, in the emergence of the G20, which has, as an informal catalyser of international agreements, largely condemned to growing irrelevance the traditional industrialized-country-geared configurations under the G7 and G8. Rio+20, and the dynamics behind a new global development agenda beyond 2015, are just two major international processes that signal the fact that the geopolitical dynamics have simply become more complex and dynamic.

The capacity of emerging countries to shape in their own terms the course and content of major multidimensional and multi-stakeholder processes – as was the case for Rio+20 – cannot be fully rationalized other than by understanding the growing national geopolitical and economic weight of these new actors. However, the growing economic and political importance does not immediately translate into influence in multilateral processes within the United Nations. Many other factors come into play, ranging from individual personalities to national strategies, when one is assessing the impact and influence on UN processes of such emerging countries.

The financial meltdown of 2008 facilitated a growing political assertiveness by specific emerging countries and the creation of new political associations which had not played a protagonist role in shaping the sustainable development agenda, particularly at the Millennium Summit of 2000, the World Summit on Sustainable Development (WSSD) of 2002 or the UNCED of 1992. Such international assertiveness was then translated into influence in the Rio+20 process through the active engagement of specific individuals, whose contributions this book attempts to bring to light.

The collective role undertaken by Brazil, Russia, India, China and South Africa – commonly referred to as BRICS, but also known in climate change negotiations as the BASIC countries, which excludes Russia as not being part of the Group of 77 – is illustrative of the drive by emerging countries to redefine, in new and more advantageous terms, the paradigm of sustainable development. However, the BRICS countries are far from being the only emerging players. From a completely different perspective, the financial crisis of 2008 also provided political momentum for the Bolivarian Alliance for the

Peoples of Our America (Spanish: Alianza Bolivariana para los Pueblos de Nuestra América, also known as ALBA), spearheaded by Venezuela, Cuba, Bolivia and Nicaragua, to assert these countries' national interests and thereby disseminate an alternative vision concerning sustainable development. Other emerging countries, such as Indonesia, the Republic of Korea, Kenya, Mexico and Turkey, have also risen in their own right as assertive central players in the definition of a new sustainable development agenda.

During Rio+20, many of these emerging developing countries did not perform a visible role in the negotiations. Therefore, it is difficult to qualify their contribution solely in terms of the initiatives presented during the course of the negotiating process, many of which were introduced under the umbrella of the Group of 77. Even so, it is important to highlight the general orientation of some of these countries in the years after the crisis and their influence in reshaping the concept of sustainable development.

The emerging role of developing countries, particularly after the financial crisis of 2008, cannot overshadow the undeniable reality that the crisis also brought into the picture a panoply of other traditional and non-traditional stakeholders, as illustrated by the origin of the overwhelming majority of voluntary pledges made at Rio+20. These covered a wide array of global corporations based in emerging markets, as well as charities and philanthropic associations, investment banks, non-governmental organizations, trade unions, education institutions, women's organizations, sub-national and local authorities, and many others frustrated by the lack of delivery of previous sustainable development agreements and therefore interested in strengthening sustainability.

The world after 2008 was also a completely different place for traditional donors, who had, over the years, grown accustomed to enjoying the lion's share in the global development agenda. Several trends forced donor countries such as the United States, Canada, Switzerland, Japan, the EU countries and Norway to review their own approaches towards sustainable development.

The first trend was a reassessment of the importance of traditional ODA. The significance of ODA dramatically diminished compared to other types of international flows, such as private investment flows and other "non-traditional sources of financing for development" between developed and developing countries and, most importantly, among developing countries. The second trend was the emphasis given after 2008 to financing climate change mitigation and adaptation by geographically reallocating the scarce ODA towards highly vulnerable LDCs, to the detriment of middle-income countries. Finally, institutional transformations, particularly among European countries (traditional leaders in advancing the global sustainable development agenda), played their part.

The coming into force of the Treaty of Lisbon in December 2009 profoundly rebalanced the coordination processes within the European Union, which so far had played a traditionally leading role in driving forward the planet-centred sustainable development agenda. The conciliation of national interests was, after Lisbon, increasingly driven by the dynamics within the European Commission, whose authority was reinforced by the establishment of a new European foreign service (the European External Action Service) and by a single legal personality for the European Union, whose role was enhanced in the United Nations after the adoption in 2010 of Resolution 65/276 on the participation of the European Union in the work of the United Nations (United Nations, 2011a). The specific role of these key developed countries will be discussed in Chapter 3.

### *Brazil, Russia, India, China and South Africa (the BRICS association)*

The resilience of some emerging economies to the impact of the economic crisis provided a political space for the BRICS association to emerge as a central player in key economic, financial and environmental multilateral processes. In June 2009, in their first meeting, held in Yekaterinburg, Russia, the leaders of the five countries agreed the objective of pursuing "key common interests from their perspective as emerging market economies and developing countries." The majority of existing analyses on the role of the BRICS countries have focused on their collective weight in the global economy and on their divergent geopolitical interests. However, the BRICS countries have not shied away from devoting a prominent role to international development issues in their common agenda. Since 2009, despite their diverging national interests the five countries have focused their attention on defining, from an essentially pragmatic perspective, broad common positions in four of the issues that will redefine the sustainable development agenda. The first issue is a common approach to sustainable development linked to the "potential contribution that emerging economies can have in driving global prosperity." The second is an overarching interest in the linkages between energy, development and climate change. The third issue is a common drive towards the reform of the international institutional architecture, particularly as regards the IMF, the World Bank and, to a lesser extent, the UN organs, with a focus for the latter on the Security Council, the General Assembly and other development-oriented bodies such as the CSD or UNEP. A fourth issue is the common defensive approach towards common goods and planetary boundaries.

The BRICS countries share the view that the implementation of sustainable development "should be a major vector in the change of paradigm of economic development" – in other words, sustainable development as "the main paradigm in environmental issues, as well as in economic and social strategies." In 2012, in the wake of Rio+20, the BRICS countries highlighted "accelerating growth and sustainable development, along with food, and energy security," as their most important common challenges. For these countries, the Rio+20 Conference was a unique opportunity for the international community to support the "overarching sustainable development framework." While the BRICS countries have always felt comfortable in accepting that sustainable development should be the main development paradigm, they have also felt compelled to emphasize a strong sense of national ownership towards all policy measures. Their 2012 position on "the green economy" is illustrative of this point. For BRICS leaders, the green economy was supposed to be a means to achieve fundamental and overriding priorities in a larger framework of sustainable development and poverty eradication. Their Delhi Declaration expressed the position that

> national authorities must be given the flexibility and policy space to make their own choices out of a broad menu of options and define their paths towards sustainable development based on the country's stage of development, national strategies, circumstances and priorities.
>
> (BRICS, 2012)

Climate change and energy security are also key common interests for the BRICS countries and are strongly perceived as interlinked developmental challenges. Although the five countries are positioned differently in their status as energy producers, consumers and transit states, their two complementary concerns have been, first, to diversify energy supply and promote energy infrastructure, and second, to diversify their own energy mix and gradually develop cleaner, more affordable and sustainable energy systems in which biofuels play an important role as environmentally friendly technologies for diversifying the energy mix and creating jobs. Climate change positions adopted particularly by the four BASIC countries largely rest on the principles of equity and of common but differentiated responsibilities and respective capabilities. In this regard, their positions usually reflect a constant balance between the need to combine measures to protect the climate with steps to carry out their "socio-economic development tasks."

Tariq Banuri, in his capacity as director of the Rio+20 Secretariat, had been arguing that the conference should put energy at the centre of its agenda. He went on to argue that if the costs of renewable energy could be driven down and its deployment ramped up in developing countries, it could end energy poverty, contribute to economic recovery and growth, generate employment in all countries and help avoid dangerous climate change (Banuri *et al.*, 2009).

International institutional reform and the governance deficits in international institutions have been a coalescing factor behind the BRICS association. The financial crisis catalysed a process – largely driven by BRICS countries such as Brazil, India and China, joined by others such as Argentina, Mexico and Indonesia – to develop changes in international financial institutions in order to reflect their relative weight as emerging markets in the global economy. At the same time, the BRICS countries, although divided by different interests, have traditionally valued the informal policy-coordinating role of the G20. In 2011, BRICS leaders called on the G20 to play a more active role in global economic governance "as the premier forum for international economic cooperation." The recourse to the G20 has in no way undermined the constant emphasis given by the BRICS countries to a "comprehensive reform of the UN" (BRICS, 2011) and in particular of its Security Council, with a view to making the peace and security pillar of the United Nations more "effective, efficient and representative so that it can deal with today's global challenges more successfully" (ibid.).

### *The Bolivarian Alliance for the Peoples of Our America (ALBA)*

ALBA is composed of Antigua and Barbuda, Bolivia, Cuba, Dominica, Ecuador, Nicaragua, Saint Vincent and the Grenadines, and Venezuela. The alliance is rooted in an ambitious political vision of its forefathers, the late president of Venezuela Hugo Chávez Frías, and the former president of Cuba Fidel Castro, of building a sub-regional political and ideological alliance capable of playing a balancing role in an international multipolar setting.

Throughout its first decade of existence, different political and ideological strains relevant to the sustainable development agenda have converged in the aspirations of the ALBA. A first strain is rooted in the nineteenth-century integrationist aspiration expressed by the Bolivarian dream of uniting the republics of the Americas. The second one draws from the reaffirmation – widely shared by all developing countries – of the Rio principle II, which stipulates that

> States have, in accordance with the Charter of the United Nations and the principles of international law, the sovereign right to exploit their own resources pursuant to their own environmental and developmental policies, and the responsibility to ensure that activities within their jurisdiction or control do not cause damage to the environment of other States or of areas beyond the limits of national jurisdiction.
>
> (United Nations, 1972)

The third strain is the embrace in its political positions, particularly after the accession of Ecuador and Bolivia to the Alliance, of indigenous peoples' cultures, spiritual traditions, histories and philosophies, especially their rights to their lands, territories and resources, in a manner consistent with the 2007 Declaration on the Rights of Indigenous Peoples (United Nations, 2007).

These views translated in the run-up to Rio into positions usually attaching a strong value to a rights-based approach to development and also on perceiving the natural environment as a subject of rights.

Finally, ALBA is also a network of sub-regional technical, trade, energy, monetary and economic cooperation mechanisms established in the Havana agreements of December 2004 subscribed between Cuba and Venezuela. These agreements were initially developed as a reaction to the ill-fated initiative of the Free Trade Area of the Americas (FTAA).

From its beginnings, the Alliance has maintained a strong focus on the "promotion of democratization and transparency within international organizations, the United Nations and its organs" (Cuba and Venezuela, 2004). This common priority was firstly identified in the 2004 joint declaration between the Venezuelan and Cuban presidents at the time, Hugo Chávez and Fidel Castro respectively. Nevertheless, with accession to membership in the Alliance of Bolivia (2006), Nicaragua (2007) and Ecuador (2009), the group of countries became more engaged in deepening their involvement in other international development processes.

The election in the summer of 2008 of Miguel d'Escoto Brockmann – a veteran statesman, politician, community leader and priest who served for over a decade as Nicaragua's Minister for Foreign Affairs – as president of the UNGA provided a solid outreach platform through which to promote transparency and democracy within international organizations. Using the Doha Review Conference on Financing for Development of November 2009 as a stepping stone, UNGA president d'Escoto convened dialogues in October 2008 and March 2009 on the world financial and economic crisis. These dialogues were instrumental in providing a platform for the Stiglitz Commission Report and for the United Nations Conference on the World Financial and Economic Crisis and Its Impact on Development. This conference, held in June 2009 in New York, was the culmination of the outreach efforts undertaken under the presidency of Father d'Escoto. The case for reassessing the role of the United Nations in global economic governance, in conjunction with the lack of response from the international financial institutions, galvanized broad support from a majority of developing countries reunited within the Group of 77. The assimilation of universality, legitimacy and consensus as three necessary preconditions to economic governance also influenced the thinking and eventually the agreements on the institutional arrangements for sustainable development during Rio+20.

Although the d'Escoto presidency touched upon many other development challenges, such as the food crisis and the right to food, the decisive move of ALBA towards sustainable development started in 2009. The seventh summit of the Alliance, held in October of that year in Cochabamba, Bolivia, was key for broadening the scope of the Alliance's agenda and for shaping its growing influence in the climate change and the Rio+20 processes. The Cochabamba summit was held a month before the climate conference of Copenhagen. It was also the venue of a broad social summit organized by the Alliance's "social movements council" – itself an "anti-neoliberal" and "anti-imperialist" space.

The timing of the Cochabamba summit and its main outcomes are important elements for understanding the influence of ALBA in the negotiation of "The Future We Want" in Rio+20. The "Fundamental Principles of the Trade Agreement of the Peoples" and the "Declaration on the Rights of Mother Earth" agreed by the presidents of member countries of the Alliance define sustainable development mainly in terms of the realization of basic human rights, such as the right to water and sanitation, the right to food, the right to education and health. Both documents are also critical in understanding the rationale of why the most active countries of the Alliance also sought to adopt the human rights-based approach towards development and apply it in a broader sense to nature through the "promotion of the rights of Mother Earth and the promotion of an economic development in harmony with nature."

Another common interest of ALBA has been its members' strong emphasis on the role of the state and their systematic opposition to the privatization of any "basic services such as the access to water, health, education, telecommunications, transportation and energy." It is therefore grounded in the principles of the Cochabamba Declaration, which is why Bolivia was able to gain support from its partners in defining, from the very beginning of the Rio+20 process, a defensive stance on the valuation of ecosystem services – which is in itself one of the core tools for any green economy framework – by successfully advocating the recognition of the different values of biodiversity (social, cultural, economic, etc.) and consistently cautioning against the "patenting of biodiversity."

The early definition of ALBA's common positions on key sustainable development issues – such as the green economy, a rights-based approach to development, and a drive for the universalization of the sustainable development institutions – contributed to its permeation, during the early part of the Rio+20 preparatory process, of the initial position of the Group of 77. In addition, the Alliance's positions were greatly enhanced by the diligence of several of its negotiators – from Cuba, Bolivia, Ecuador, Nicaragua and Venezuela – who were fundamental in shaping the final sections of the Rio+20 outcome document, "The Future We Want." The content of paragraph 39 of "The Future We Want," which recognizes that "planet Earth and its ecosystems are our home and that 'Mother Earth' is a common expression in a number of countries and regions, and [notes] that some countries recognize the rights of nature" (United Nations, 2012b), as well as paragraph 56 at the opening of the green economy section, which affirms that "there are different approaches, visions, models and tools available to each country, in accordance with its national circumstances and priorities, to achieve sustainable development" (ibid.), were landmark achievements for the ALBA countries.

### *Indonesia and ASEAN*

In 2010, Indonesia assumed the chairmanship of the Association of Southeast Asian Nations (ASEAN), which comprises ten of the most dynamic emerging economies (Brunei, Cambodia, Indonesia, the Lao People's Democratic Republic, Malaysia, Myanmar, the Philippines, Singapore, Thailand and Vietnam) and close to 9 per cent of the world's population. This position allowed the world's fourth most populous nation to consolidate its leadership among developing countries on the basis of its sustained economic growth, solid governance and a widely respected diplomatic tradition. Indonesia's engagement in the multilateral process after the financial crisis attempted to transcend the prevailing conflicting political dynamics. Its overall take on the political situation of 2008 was that the crisis provided an opportunity for developing countries to take the lead in shaping a global sustainable development agenda. Indonesia's positioning in the United Nations following the outbreak of the financial and economic crisis was shaped by a series of distinguished ambassadors – such as Marty M. Natalegawa, who was promoted Minister of Foreign Affairs; Hasan Kleib, who later served as Special Presidential Envoy for the UN Post-2015 Development Agenda; and Desra Percaya – and their teams of sustainable development specialists.

For Indonesia's decision-makers, all countries were going to be judged by how they measured up against global sustainability challenges. Building on its leadership in ASEAN and during the 2010 Cancún, 2011 Durban and 2012 Doha climate conferences, as well as the Rio+20 process, Indonesia advocated within the Group of 77 that both developed and developing nations must do more and move away from the "business as usual" mentality. Indonesia promoted the idea that while developed countries must take the lead, because of their historical responsibilities, and fulfil their commitments with more ambitious targets, developing countries must also seriously play their part by contributing concretely to resolving global challenges, including through thinking outside the box and moving beyond long-standing entrenched positions. In the view of Indonesia's leaders, only through the embrace of their respective responsibilities would it be possible for developed and developing countries to ensure actionable outcomes in climate change, achieve the MDGs and make Rio+20 a turning point in the sustainable development paradigm. It was also clear from Indonesia's own experience that the actions of developing countries had to be framed within the principle of common but differentiated responsibilities and respective capabilities.

A country composed of over 17,000 islands had its own clear priorities with regard to sustainable development. Clearly, many of these were related to its own national drivers of economic growth, such as management of agriculture and forestry, food industries and water resources management. However, its privileged political positioning within the United Nations allowed Indonesia to play an often irreplaceable convening role on several other key areas, such as the institutional framework for sustainable development, climate change, biodiversity, ecosystem valuation, and sustainable consumption and production patterns. Indonesia's pragmatic ambition was recognized with the appointment of President Susilo Bambang Yudhoyono to lead a high-level panel of eminent personalities – jointly with the prime minister of the United Kingdom, David Cameron, and the president of Liberia, Ellen Johnson Sirleaf – and shape the first outline of the UN Post-2015 Development Agenda.

### *The Republic of Korea*

The Republic of Korea is a country that has made, in the course of a lifespan, the transition from being one that receives aid to being one that gives aid and from a country whose recovery was helped by the United Nations to one that leads within the organization. In 1992, Korea was still a developing country and a member of the Group of 77. In 1996 it became a member of the Organisation for Economic Co-operation and Development (OECD). In 2001 the then Minister of Foreign Affairs, Dr Han Seung-soo, was elected president of the UNGA. Six years later, Ambassador Ban Ki-moon, who had served as chief of staff in Dr Han's presidency, took office as the eighth Secretary-General of the United Nations. In June 2011, Ban Ki-moon was re-elected for a second term under a pledge to make sustainable development an overarching priority of the United Nations. It is also not without significance that in 2012 another Korean native, Jim Yong Kim, a medical doctor and PhD, became the president of the World Bank, propelling Korea into the leadership of two of the most powerful multilateral and global institutions.

In times of drastic economic and financial change, Korea also offered to lead the G20. This country was – as was pointed out in 2010 by President Lee Myung-bak – the first country from outside the ranks of the G8 to chair the G20. After taking the reins of the G20 from Canada and the UK, and despite resistance from other partners, Korea pushed to address the crisis in a way that did not solely reflect the perspective of the financial markets. The scope given to the G20 by the Korean presidency shifted from a focus on immediate and short-term measures of financial reform and stabilization to initiate, among world leaders, a different conversation on the longer-term policy coordination framework required to ensure global prosperity. It is in this context that within the G20 there emerged the idea of considering long-term strategies to achieve balanced and sustainable growth and close the gaps between advanced and developing countries. By addressing broader developmental challenges as part of the G20 leaders' summit agenda, the Korean presidency contributed to better defining the role of the G20 as an informal policy coordination platform that can catalyse supportive action for goals set in the context of the United Nations – thus reinforcing a supportive, albeit à la carte, relationship between the G20 and the United Nations, particularly throughout the Korean (2010), French (2011), Mexican (2012) and Russian (2013) presidencies of the G20.

The emerging role of Korea as a leader in the promotion of sustainable development, particularly through low-carbon development and green growth strategies, was further embodied by its bid to host the 2012 climate change conference and by the election of In-Kook Park, ambassador to the United Nations, to preside over the governing bureau of the Rio+20 preparatory process – a responsibility that Ambassador Kim Sook took over in July 2011. The deputy ambassador of Korea, Boonam Shin, also had a key role in the seminal work leading to the decision to convene Rio+20 and in the choice of the green economy as one of the themes for the conference. Boonam joined the mission after presiding over the Environmental Policy Committee of the OECD and in 2011 was appointed special Ambassador for Green Growth.

## Mexico

Mexico enjoys a very special position within the United Nations. It is a developing country that is not a member of the Group of 77. This unique situation allows it to play a bridging role in the reconfiguration of the development pillar of the organization. In 2002, Mexico fostered the realization of the Monterrey Consensus, which today remains the basis of the means of implementation for the MDGs and of the financing for the development agenda. In 2005, in the wake of the World Summit, Mexico spearheaded the reform in the human rights pillar of the United Nations that led to the establishment of the Human Rights Council. The following year, jointly with Switzerland, Mexico led the process of reform of international environmental governance, which concluded after Rio+20 with the revamping of UNEP and the establishment of the United Nations Environment Assembly.

In late 2008 Mexico decided to host in 2010 the Sixteenth Conference of the Parties to the UN Framework Convention on Climate Change and the Kyoto Protocol (UNFCCC COP16/MOP6). At that time, the expectations for that meeting were uncertain, since the international attention was focused on reaching a comprehensive agreement at the Fifteenth Conference of the Parties in 2009 in Copenhagen, the latter upgraded to the level of summit. After the failure of the summit, Mexico's role and leadership in the shaping of the sustainable development agenda was substantially enhanced by three fundamental contributions. The first one was by the inclusive process of outreach and consultations. The Mexican efforts helped to restore confidence in the UN and in multilateral processes related to sustainable development during the critical juncture accelerated by the economic crisis and the emergence of the G20. The second contribution to the sustainable development agenda are the Cancún Agreements, particularly the universal recognition of the climate threshold of 2 degrees in the increase in the global temperature. The third contribution was the reconfiguration of the financing for climate change architecture through the establishment of the Green Climate Fund.

In 2011, Mexico, along with other engaged countries such as Indonesia and Kenya, initiated a discussion on the lack of coherence among the institutions responsible for setting and reviewing the development policy of UNEP. This discussion then provided the platform at Rio+20 for replacing the CSD and enhancing the agenda-setting and review functions in the field of sustainable development of both the United Nations General Assembly and the Economic and Social Council through the establishment of a High-Level Political Forum.

Before the 2008 crisis, Mexico joined other countries that were pushing for a broad reform of the international financial architecture and for enhancing the voice of developing countries in global governance. The Group of 5 (G5) – a dialogue platform established in 2005 involving the leaders of Brazil, China, India, Mexico and South Africa – played a seminal role in broadening global policy coordination and in catalysing a transition from the G8 setting to the

G20. Mexico also emerged as a proponent of this transformation of the global policy coordination framework.

Like other developing countries, Mexico was also a vocal critic of the inability of the IMF and the G8 countries to foresee and in a timely way undertake many of the economic, fiscal, financial and trade reforms introduced – not without painful social consequences – in several emerging countries during the 1990s. For developing-country G20 leaders, the financial turmoil in the United States and Europe was above all a reflection of a collective failure to transpose decisions into regulation in an early and effective manner.

Mexico's case for reform was first articulated by its own bid in 2011 to take the leadership of the IMF through the governor of its central bank. The Mexican view was that the international institutional framework, starting with (but not limited to) the IMF and the other multilateral financing and trade institutions, was not adequately performing several of its fundamental functions (Carstens, 2011). In Mexico's view the glaring institutional deficiencies derived primarily from the incapacity of international institutions to adapt to a changing and evolving international landscape. The under-representation of emerging developing countries in the governance of multilateral financing and trade institutions was only one of the symptoms of this deficit. Another symptom was the inability of multilateral institutions to open up and involve other non-state actors in their processes. A third systemic deficiency was the lack of institutional competence and the lack of institutional surveillance – whether financial, environmental or political – and the necessary flexibility to prevent and resolve crises. The fourth, and perhaps most important, deficiency identified by Mexican leaders and acutely exposed by the failed response of multilateral institutions to the financial crisis was their increasing inability to induce, at the right time and the right level, adequate policy coordination.

Mexico has always believed that international policy coordination is essential but extremely difficult to engineer. On the basis of these shortcomings, Mexican leaders identified a valuable opportunity to use the presidency of the G20 to reinforce international institutional governance and policy coordination. In this regard, Mexico followed the path set by the Korean presidency and undertook the broadest stakeholder outreach and consultations effort ever yet undertaken within the G20.

## *The African Union and Its East African Community*

The emergence since the first Rio Summit of new regional or sub-regional associations (as has been illustrated by the cases of ASEAN and ALBA) has contributed to transforming the political dynamics that shape the sustainable development multilateral process. Among the most remarkable changes in this regard was the evolution in 2002 of the Organization of the African Unity towards the African Union. Such transformations run parallel to the creation of sub-regional associations, such as, in 1992, the Southern African Development

Community (SADC), which unites 15 Southern African countries. This has been replicated in the East African Community (EAC), which was established in 2000 by Kenya, Tanzania and Uganda, later joined by Burundi and Rwanda, transforming the region into an economic, commercial and political hub for Eastern Africa and the Great Lakes Region. The five EAC countries have also emerged in the immediate past as invaluable players in the reshaping of the sustainable development agenda. They cover regions that are among the most vulnerable to climate change and yet retain – along with the Congo Basin – the world's second-largest biodiversity pool. Despite their development challenges, the EAC and SADC countries have experienced during the past decade some of the world's highest sustained rates of economic growth and have been undeniably transformed by the impact of the MDGs. Because of these factors, and especially as a result of the strengthening of the political and coordination role of the African Union, the Eastern African countries have been in a privileged position to contribute to shaping an ambitious African position and to deeply influence the negotiating dynamics within the Group of 77.

Spearheaded by Kenyan leaders and Kenya's representatives in the United Nations, the priorities of the East African countries with regard to some of the key issues included in the Rio+20 process can be illustrated, first, through their approach to the green economy as a potential source for accelerating their national economic transformation, provided it is not interpreted as a limiting framework.

The African common position towards the green economy – heavily influenced by the EAC and SADC countries – was based on the realization that African countries, being at the early stages of industrialization, have "an opportunity to pursue sustainable industrial growth that limits the environmental, social and economic costs of industrialization, and increases the efficient use of energy and material input, thereby enhancing international competitiveness" (UNECA, 2011). The second common priority was to revalue and consolidate the UN institutional presence in Africa "and thus to strengthen, consolidate and transform UNEP into an international specialized institution for the environment based in Nairobi, Kenya." A third priority for African countries was to reassess the important extent of their natural wealth by developing "new reference indicators to assess the economic, social and environmental performance of their economies, alongside the Gross Domestic Product" and adopting policies that promote the integration of the true environmental costs of production and consumption into accounting models. A last common African priority was retaining the international commitment towards the accelerated achievement of MDGs. As a consequence of this last priority, African countries took a defensive stance towards new international Sustainable Development Goals in isolation from the MDGs, which eventually forced the so-called development and the sustainable development communities to push for a convergent integrated development framework.

### *The Caribbean Small Island Developing States*

Rio+20 was the first multilateral conference in which the Small Island Developing States (SIDS) of the Caribbean played a prominent role and had citizens at every level of leadership. Ambassador John W. Ashe of Antigua and Barbuda was co-chair of the Bureau and brought considerable experience and leadership in development processes to the task. It was Barbados, however, that played a seminal role. Its prime minister, David Thompson, who died during the process, served on the Secretary-General's High-Level Global Sustainability Panel. He was later replaced by Thompson's successor, Prime Minister Freundel Stuart, who served out the rest of the Panel's term. Another Barbadian national, in the person of Elizabeth Thompson, served as a UN Assistant Secretary-General and executive coordinator of the Rio+20 Conference. She was a former minister who, before joining the United Nations, had led the formulation of the Latin American and Caribbean region's first national green economy policy. Barbadian negotiator Selwin Hart, as a leader of one of the critical negotiating splinter groups set up by the co-chairs, was instrumental in keeping negotiators at the table in some of the more divisive issues when fracture threatened to derail the negotiation process at Rio. Barbados also played a critical role in the work of the "groups of friends" (see Chapter 2), particularly the friends of the seas and oceans. In May 2012 it was in Barbados, the birthplace of the SIDS Barbados Programme of Action (BPOA) of 1994, which defined the international agenda for Small Island Developing States, that the SIDS launch of the Sustainable Energy for All (SE4All) initiative and the Rio+20 review took place. That meeting was attended by representatives at the highest political level and was the first regional meeting, prior to the conference, to commit explicitly to SE4All and the green economy as tools for poverty eradication and sustainable development.

## The science underpinning an integrated approach towards development

Since Stockholm, a number of environmental reports have attracted the attention of the media, governments and varied stakeholders, with some becoming respected international staples. Prior to the 2012 Rio+20 Conference, one report of particular influence was the *Global Environment Outlook 5* (*GEO-5*) report by UNEP (2012). This report gave a snapshot of the state of the planet and the responses of member states to environmental and development issues. It also provided a critical insight into the thematic areas to be considered at Rio+20 and in some respects offered a backdrop to the conference, which underscored the importance and urgency of the issues to be addressed by policy-makers at Rio.

The Stockholm Conference Declaration states:

> Man is both creature and moulder of his environment . . . In the long and tortuous evolution of the human race on this planet a stage has been

> reached when, through the rapid acceleration of science and technology, man has acquired the power to transform his environment in countless ways and on an unprecedented scale.
>
> (United Nations, 1972)

Forty years later, these sentiments would again be voiced by scientists who indicate that we are living in a new era, the Anthropocene, characterized by man's influence on the natural resources and the planet "on a scale akin to the great forces of nature" (Anthropocene, 2012).

## *The Fifth* **Global Environment Outlook** *report (GEO-5)*

The *GEO-5* report gave strong evidence that now, more than ever, the words penned in the Stockholm Declaration are prophetic, and that man as "moulder" of his environment has an exponential impact on earth's ecosystems: "the power to transform his environment in countless ways and on an unprecedented scale" (UNEP, 2012).

*GEO-5*'s "Summary for Policy Makers," a negotiated document by member states, was ready for UNEP's Governing Council in February 2012. As part of the rigour, integrity in method, standards-setting and production process of *GEO-5*, UNEP brought together a High-Level Intergovernmental Advisory Panel, a Science and Policy Advisory Board and a Data and Indicators Working Group. Brazil had played a critical role in the development of the outline for the 2012 report, with one of its government experts being active on the Advisory Panel. The summary was endorsed by a number of governments, including those of Brazil, Canada, the Czech Republic, Egypt, the Republic of Korea, Pakistan and the United States – all members of the Rio+20 Bureau. *GEO-5* was launched in Rio by the executive director of UNEP, Achim Steiner, on the eve of the high-level segment of the Rio+20 Conference (UNEP, 2012). The report was an important contribution to Rio+20 and built on previous *GEO* reports, which had provided an analysis of the state of trends and an outlook concerning the global environment. As a result of this process and structure for the production of the Report, UNEP asserts that *GEO-5* is "a comprehensive, integrated and scientifically credible global environment assessment," with "policy-relevant information, including the analysis of cross-cutting issues and indicator-based components" (ibid.).

The report (UNEP, 2012) covers a number of very critical issues:

- acceptance of critical thresholds;
- the need for evidence-based policy-making, which means the need for more reliable data;
- a shifting policy focus towards the drivers that contribute to an increased pressure on the environment (e.g. globalization, population growth, consumption and production and urbanization);

- the scaling up of promising policies in areas such as fresh water, biodiversity, climate change, land, chemicals and waste, energy, oceans and seas and environmental governance;
- innovative responses to improve human well-being;
- recognition that environmental deterioration demonstrates that internationally agreed goals have been only partially achieved.

The last point is very significant as it shows that environmental goals alone were not being delivered, but when linked with those in the social and economic arena they were much more likely to be achieved.

Just before the start of the Rio+20 High-Level segment, *GEO-5* set a tone of urgency for the conference and member states. It was a sobering document. One wonders, however, whether the timing of the launch militated against *GEO-5*'s receiving further attention at Rio and catalysing the kind of momentum that might have taken place had it been launched even six months earlier. At the time of the launch, in June 2012, the negotiations were already well under way. Negotiators and arriving ministers and heads of state would not have had sufficient time to digest the full report and make changes to their submissions based on its contents. In fact, they were intensely focused on having an agreed outcome text, one practical and capable of implementation. Moreover, at Rio the opportunities for the addition of significant new text had been considerably reduced by the protracted third round of informal-informal[3] negotiations that took place from May 29, 2012 to June 2, 2012 in New York. During that last negotiation round before the start of the conference, there was already high pressure to get member states to work with the existing draft text with a view to coming to a consensus rather than producing new drafts or introducing new amendments to the text.

*GEO-5* was also of significance given the three panels that were used to ensure the integrity, rigour and the quality of the report. It was filled with scientific and technical data collected using stringent methodologies. In this context, *GEO-5* served to satisfy and answer one of the recurring calls that took place in the lead-up to, during and after the conference: the need for sound science-based data on which policy-makers could make more informed decisions and which would help them understand the climate and circumstances within which those decisions must be made.

In multilateral negotiations, language and nuance assume tremendous importance. Rio+20 was no exception, and there had been debate and divide around the use of the expressions "critical thresholds" and "planetary boundaries" to describe those natural resource markers and losses up against which human activities bring us and which might cause irreparable harm if left unmitigated. *GEO-5* came down in favour of the term "critical thresholds," which had also been favoured by the host country, Brazil.

Despite the less than ideal timing of its launch, *GEO-5* also powerfully paints the background of the canvas on which the Post-2015 Development

Agenda must be overlaid. In this regard it is a useful and necessary report on the "state of the planet" which underscored the necessity of the Rio+20 Conference and its decisions. In particular, *GEO-5* emphasized the need for integrated policy approaches to development; the importance of stakeholders as partners in implementation; the inescapable role of the private sector as a major creator of jobs, innovator, developer of technology, provider of capital and consumer of natural resources; and the necessity for a global green economy with specific national characteristics. The green economy as pictured in *GEO-5* understands that growth is still needed, while at the same time it is essential to build human and social capital and to protect the natural resources on which development gains, economic and social growth are predicated.

Finally, *GEO-5* is an invaluable reference guide for current and aspiring policy-makers, technocrats and the development institutions that serve them as to why and how they can and must implement policy in the Anthropocene era. It also provides the science and empirical data for shaping the policy positions necessary to effect the "transformative change" to which *GEO-5* speaks.

### *Synthesis report on best practices and lessons learned on the objective and themes of the Rio+20 Conference*

Another very important report produced during the Rio+20 process was the "Synthesis report on best practices and lessons learned on the objective and themes of the United Nations Conference on Sustainable Development" (SR), which was published in two versions (United Nations, 2011b). The first was released in January 2011 to coincide with the First Intersessional[4] in the process leading up to the conference. The second, revised publication was in March 2011 to coincide with the meeting of the Second Preparatory Committee (PrepCom).[5] The report was framed around the objectives and themes of Rio+20, with a section detailing the way forward. The SR was compiled by initially sending questionnaires to member states to solicit their experiences and views on the issues revolving around the objectives and themes of the Rio+20 Conference, including risks, challenges and successes in policy approaches. All member states were sent the questionnaire, as well as the Major Groups,[6] international financial institutions, and agencies within the UN system. Some 108 responses were received, comprising 49 from member states (including the European Union), 32 from major groups and 27 from UN agencies. In describing the purpose of the report and the function it was expected to fulfil, the drafters stated that "the present report seeks to synthesize the main points in an objective, balanced and coherent manner based on replies to the questionnaire and other statements, as appropriate."

In the report, member states emphasized the high level of political ownership and support for the sustainable development agenda, as was demonstrated by greater information-sharing, increased transparency, new legislation and policy, gender mainstreaming, increased budgetary allocations, increased support for stakeholder participation, institutional development and strengthening, as well as the execution of development activities. The agencies of the UN system attached great importance to sustainable development frameworks being "inclusive, holistic, balanced and integrated." The report also registered a "shift in the development dialogue" from infrastructural issues, such as transport and housing, to climate change, carbon reductions by industry, support for policy and technologies focused on energy efficiency and renewable energy, catastrophe planning and disaster risk reduction. It was the general view that integrated decision-making, while necessary and important, had been a "low political priority." Nonetheless, the need for the formulation and implementation of national sustainable development strategies and inter-ministerial coordinating mechanisms was also regarded as a possible tool to advance sustainable development. In addition, the following ideas were noted: policies and strategies relating to green growth, green procurement, green investing, green taxes, eco-labelling, increased social entrepreneurship, and greater acceptance and use of the concept of corporate social responsibility (CSR). Member states were urged to note that, in the transition to the global green economy, biodiversity has to be protected and education for sustainable development has to be mainstreamed. Some countries suggested that the creation of a platform for sharing best practices and lessons learned could be useful in avoiding pitfalls and maximizing opportunities and benefits from success stories.

Issues such as ODA, the role of the developed countries in providing funding, debt cancellation, the removal of trade barriers, as well as building support for technology transfer and investment in institutional and technical capacity, were all highly rated as central to giving momentum to the development process in the developing world. Small Island Developing States (SIDS), LDCs and women were thought of as being worthy of special consideration and support; while jobs for youth were seen as an increasingly important issue which also needed to be addressed.

New and emerging challenges were identified: climate change; related natural disasters; the interrelated financial, economic and food crises; energy security; the degradation of ecosystems, in particular marine ecosystems; and diminishing natural resources, including water scarcity. Additional challenges identified in the *GEO-5* report as drivers were unsustainable consumption and production, the impacts of population growth, rapid urbanization, political instability and social unrest.

While it was generally agreed that the green economy had value as a mechanism for poverty eradication and sustainable development in all countries, both developed and developing ones, it was considered that there could be no "one size fits all" approach; that green protectionism should be guarded against and avoided; and that green taxes should be implemented, although subsidies were seen as harmful. It was also felt that an appropriate enabling environment had to be created to facilitate the transition from the brown to the green economy through appropriate fiscal incentives, legislative and regulatory frameworks, procurement strategies and a shifting focus to energy efficiency and renewable energy technologies.

At the institutional level, the future of the CSD, a broadened role for ECOSOC, the strengthening of UNEP, the creation of "an umbrella organization/structure," and the scope and potential of "the UN delivering as one" were all raised, although it was the support for a Post-2015 Development Agenda with MDGs that would become more focused on as a sustainable development objective. The role of local governments in assisting in delivering a development agenda was highlighted, as were South–South and North–South knowledge-sharing.

The SR would later form the foundation of the "Zero Draft" on which negotiations and the final Rio+20 Outcome Document would be based. The Zero Draft was presented by the co-chairs – the ambassadors of Antigua and Barbuda and of South Korea – to member states in January 2012. At that point the SR had been available for a year. The Zero Draft eventually went through several iterations based on submissions from member states, development institutions, the private sector, the Major Groups and other stakeholders. Immediately after the Zero Draft was issued, there was a flood of criticism about how underwhelming and unambitious a document it was. Fingers pointed at the co-chairs and the Secretariat, but no country, or for that matter stakeholder group, pointed a finger at itself. This was particularly interesting since out of the United Nations' 194 member states, a mere 25 per cent responded to the questionnaire sent out by United Nations Department for Economic and Social Affairs (UN DESA). This questionnaire therefore became of extreme importance in the Rio+20 preparatory process.

The question is: did Rio+20 appear too far in the future to fully capture the attention of governments, which were still consumed by the fight with spiralling food and fuel costs against dwindling financial resources, and with citizens who were protesting their plight in the wake of the financial crisis and ensuing social meltdown and who were holding their national politicians directly responsible? The major governments that faced elections during this period, notably those of France and Germany, all fared poorly at the polls. Was it that there was insufficient coordination and prodding by UN missions in New York to get their capitals to provide answers? Could it be that the very fractured and diverse nature of sustainable development – spanning ministries, crossing the gamut of issues and subjects in society, the economy and the environment – really did not have the structures in place for national cross-cutting dialogues? And that the Rio+20 agenda therefore became a victim of a non-embedded sustainable development agenda and a bureaucracy that did not have adequate systems of communication and collaboration for framing the issues, aspirations and vision needed at Rio+20? Could it have been that there was a prevailing sense of global despair focused on survival, which removed attention from the aspirational agenda then needed and which resulted in survival winning out over vision and aspiration?

Fortunately, as the process progressed and the "Synthesis Report" evolved and gave way to the Zero Draft, the attention and activities of member states became more focused on the text, its substance, content, intent and direction.

Even against these questions, it should be noted that ultimately Rio+20 did more than confirm an existing agenda. Rio+20 broke new ground in the area of land degradation;[7] new measurements of development, beyond GDP; a strengthened role for the private sector and other stakeholders in achieving sustainable development; support for women (despite the exclusion of reproductive rights, which became a controversial issue after the Rio+20 Outcome Document was finalized); dialogue and agenda impulse on the issue of a global green economy; a possible future agreement on corporate sustainability reporting; and initial work on a new framework on sustainable consumption and production. All of these themes were to be found in the SR, later in the Zero Draft and ultimately in the Rio+20 Outcome Document. With this in mind, the criticisms of the SR may be regarded as having been too harsh.

Under-Secretary-General Sha (secretary-general of the Rio+20 Conference) reiterated that the process was to be driven by the member states, and much provision was made to ensure that this was the case, with member states being vigorous in their revisions of, responses to and comments on the Zero Draft. This was not the way in which the elaboration of Agenda 21 was approached at Rio 1992. There, "experts" worked with the secretariat to produce a high-level first text. In view of how fractious multilateral negotiations have become, perhaps it is time to consider going back to that approach.

## Notes

1 Soft law is constituted of "regional and global agreements, treaties which may not come into force, codes of practice, recommendations, guidelines, standards and declarations of principles, resolutions, that are not legally binding" but may still have some level of adherence at national, regional and international levels (Birnie and Boyle, 2002).

2 Hard law is binding and is "represented by custom, treaty, and established by general principles of law" which have been enacted (Birnie and Boyle, 2002).

3 An informal-informal meeting can be small, large, with or without structure, with or without a facilitator and with or without representatives of the Secretariat present. Unlike the usual UN meetings, in which there is continuous simultaneous extensive interpretation in the United Nations' official languages – French, Spanish, English, Russian, Chinese and English – no interpretation facilities are provided for informal-informal meetings, so they are conducted in English. Informal-informals can be used to save money.

4 Three intersessional meetings took place before the Rio+20 Conference. An intersessional meeting is where people come together to prepare for an upcoming event (United Nations, 2012a).

5 PrepCom is an "[a]bbreviation for preparatory committee (usually preparing for the creation of an international organization)" (Walker and Boyer, 2005).

6 Since the first United Nations Conference on Environment and Development in 1992 – the Earth Summit – it has been recognized that sustainable development cannot be achieved by governments alone. The nine Major Groups that aid the UN system comprise Business and Industry, Children and Youth, Farmers, Indigenous Peoples, Local Authorities, NGOs, the Scientific and Technological Community, Women, and Workers and Trade Unions (Sustainable Development Knowledge Platform, 2013).

7 Rio+20 was the first multilateral conference to agree to "a land degradation neutral world" (United Nations, 2012b).

## References


Anthropocene (2012) The Anthropocene: An Introduction. Available online at: www.anthropocene.info/en/anthropocene

Banuri, T., AtKisson, A. and O'Connor, D. (2009) A Global Green New Deal for Climate, Energy, and Development. Available online at: www.un.org/esa/dsd/resources/res_pdfs/publications/sdt_cc/cc_global_green_new_deal.pdf

Birnie, P. and Boyle, A. (2002) *International Law and the Environment*, 2nd edn, Oxford, Oxford University Press

Brazil (2008) Brazilian concept note, Brasilia, Government of Brazil. Available online at: www.stakeholderforum.org/fileadmin/files/Brazilian%20Govt%20Concept%20Paper%20Rio+20%20Nov%202008.pdf

BRICS (2011) Sanya Declaration, Beijing, BRICS. Available online at: www.brics.utoronto.ca/docs/110414-leaders.html

BRICS (2012) Fourth BRICS Summit: Delhi Declaration. Available online at: www.cyprus.mid.ru/en/pr_rel08.htm

Carstens, A. (2011) Statement to the IMF Executive Board, June 21. Available online at: www.imf.org/external/np/sec/pr/2011/pr11246.htm

Cuba and Venezuela (2004) Declaración conjunta, Venezuela–Cuba, Havana, Government of Cuba. Available online at: www.cuba.cu/gobierno/discursos/2004/esp/d141204e.html

G20 (2008) Declaration of the Summit on Financial Markets and the World Economy, Washington, G20. Available online at: http://georgewbush-whitehouse.archives.gov/news/releases/2008/11/20081115-1.html

IISD Reporting Services (2008) Second Committee Approves Draft Resolutions, New York, SIDS Policy and Practice. Online, available at: http://sids-l.iisd.org/news/second-committee-approves-draft-resolutions/

International Monetary Fund (2009) World Economic Outlook: Crisis and Recovery, Washington, DC, IMF. Available online at: www.imf.org/external/pubs/ft/weo/2009/01/

Pérez Verdía-Canales, C. (2013) Statement during the High-Level Thematic Debate of the General Assembly "The United Nations and Global Economic Governance," April 15. Available online at: http://mision.sre.gob.mx/onu/images/Disc_Gobernanza%20economicaENG_15abril13.pdf

Serra, N. and Stiglitz, J. E. (eds) (2008) *The Washington Consensus Reconsidered: Towards a New Global Governance*, Oxford, Oxford University Press

Stakeholder Forum (2008) Donostia Declaration, London, Stakeholder Forum. Available online at: www.earthsummit2012.org/about-us/milestones-to-rio20/68-san-sebastian

Stiglitz, J. (201) *The Stiglitz Report: Reforming the International Monetary and Financial System in the Wake of the Global Crisis*, New York, The New Press

Strong, M. (1992) Closing statement to the Rio Summit. Available online at: www.mauricestrong.net/index.php/speeches-remarks3/37-rio3

Summers, D. (2009) Financial Crisis: Gordon Brown Calls for New Global Order. Available online at: www.guardian.co.uk/politics/2009/jan/26/gordon-brown-economic-policy

Sustainable Development Knowledge Platform (2013) Major Groups, New York, UN. Available online at: http://sustainabledevelopment.un.org/majorgroups.html

UN News Centre (2007) President Lula of Brazil Proposes "Rio + 20" to Follow On Earth Summit. Available online at: www.un.org/apps/news/story.asp?NewsID=23952&Cr=general&Cr1=debate#.UX4Mtcote

United Nations (1972) Declaration of the United Nations Conference on the Human Environment, New York, UN. Available online at: www.unep.org/Documents.multilingual/Default.asp?DocumentID=97&ArticleID=1503

United Nations (1987) *Our Common Future*, New York, UN. Available online at: www.un-documents.net/wced-ocf.htm

United Nations (2000) The Millennium Declaration, New York, UN. Available online at: www.un.org/en/development/devagenda/millennium.shtml

United Nations (2007) United Nations Declaration on the Rights of Indigenous Peoples, New York, UN. Available online at: www.un.org/esa/socdev/unpfii/documents/DRIPS_en.pdf

United Nations (2008a) Doha Declaration, New York, UN. Available online at: www.un-ngls.org/IMG/pdf/0812_TheRoadtoDoha_Issue19.pdf

United Nations (2008b) Implementation of Agenda 21, the Programme for the Further Implementation of Agenda 21 and the outcomes of the World Summit on Sustainable Development, New York, UN. Available online at: www.stakeholder forum.org/fileadmin/files/GA_resolution_November_2008%20Earth%20Summit%202012.pdf

United Nations (2009) UN Conference on the World Financial and Economic Crisis and Its Impact on Development, New York, UN. Available online at: https://cms.unov.org/documentrepositoryindexer/MultiLanguageAlignment.bitext?DocumentID=9ad51d18-ff3e-4a0b-9d64-6c065af7b8cf&DocumentID=d09f2c69-d254-4dd4-a0dd-ab4961c68ad9

United Nations (2010) Keeping the Promise: United to Achieve the Millennium Development Goals, UNGA 65/1, New York, UN. Available online at: www.un.org/en/mdg/summit2010/pdf/ZeroDraftOutcomeDocument_31May2010rev2.pdf

United Nations (2011a) Participation of the European Union in the Work of the United Nations, New York, UN. Available online at: www.unric.org/html/english/library/backgrounders/EU.pdf

United Nations (2011b) Synthesis Report on Best Practices and Lessons Learned on the Objective and Themes of the United Nations Conference on Sustainable Development, A/CONF-216/PC/8, New York, UN. Available online at www.uncsd2012.org/index.php?page=view&type=400&nr=33&menu=45

United Nations (2012a) Rio+20: United Nations Conference on Sustainable Development. Available online at: www.uncsd2012.org/meetings_intersessionals.html

United Nations (2012b) The Future We Want, New York, UN. Available online at: https://rio20.un.org/sites/rio20.un.org/files/a-conf.216l-1_english.pdf.pdf

United Nations Economic Commission for Africa (2011) Africa Consensus Statement to Rio+20, Addis Ababa, UNECA. Available online at: www.uncsd2012.org/rio20/content/documents/325Final%20Africa%20Consensus%20Statement%20to%20Rio20.pdf

UNEP (2012) *Global Environment Outlook 5*, Nairobi, UNEP. Available online at www.unep.org/geo/geo5.asp

Walker, R. A. and Boyer, B. (2005) *A Glossary of Terms for UN Delegates*, Geneva, United Nations Institute for Training and Research

Chapter 2

# Chronicling the key impacts on Rio+20

## How Rio+20 was constructed

It is frequently assumed that processes for policy definition within the United Nations reflect global trends, national interests defined in headquarters, and positions faithfully interpreted by negotiators enjoying almost perfect information. From this perspective, the relative geopolitical or economic weight of countries is seen as the decisive factor for defining the end result of negotiations. This approach can lead to simplistic and misleading conclusions, such as the idea that major international summits fail or succeed because of the way main international powers act. However, the reality of how multilateral outcomes are shaped can trump theory.

The capacity of countries to shape the course and content of a major international conference such as Rio+20 – which was essentially a global, complex, multidimensional and multi-stakeholder agenda-setting process – cannot be fully understood using classical geopolitical variables. Several factors – some intangible, others less so – limited or enhanced the ability of countries to ultimately define the outcome of Rio+20.

A first factor is the way government representatives structure, usually one or two years in advance, the "modalities" or, more colloquially, the "rules of the game" of the negotiation. This can determine what is achievable and what is not, as much as who gets to engage and who does not.

A second factor is the importance acquired in the policy-making process by UN organizations and their secretariats, and their role as repositories of institutional memory (though some NGOs have also played this role). They have expertise in an increasingly complex web of international development processes, and an experience acquired about how to support implementation. UN secretariats are increasingly moving from merely providing administrative support to allies that, in many cases, help governments connect the dots.

A third factor is the limiting role imposed by the "consensus rule." On its positive side, this sometimes unwritten precept enhances the collective motivation and desire of all actors to achieve a final result without the need to make use of voting. On its negative side, it can lead to results that reflect lowest common denominators.

A fourth limiting factor is the gradual evolution of the process of negotiation and the length of the negotiated document itself, which forces almost all negotiators to focus on specific parts of the outcome document without necessarily seeing the whole. This relegates to a secondary level the overall consistency, construction and content of the final outcome.

A fifth set of factors are intangibles such as the leadership provided by presiding officers – usually the Bureau; the work undertaken in informal settings; the influence of Major Groups and other stakeholders, or specific issue-based interest groups such as the "groups of friends;" and, finally, the personality, technical expertise, networking skills and degree of contacts of each negotiator. Rio+20 was very much a captive and consequence of all of these dynamics.

This chapter attempts to chronicle from this perspective the major events and actors in the preparatory process of Rio+20. Between the adoption on December 24, 2009 of the United Nations General Assembly (UNGA) decision to hold a new conference on sustainable development and the official start of the conference itself, on June 20, 2012, there were a number of significant governmental and stakeholder organized meetings, events and processes at which these dynamics played out and which ultimately framed what was and was not possible.

## The formal preparations process and the lessons to be drawn

The formal preparatory and negotiation process of Rio+20 covers all those activities conducted in open meetings between May 16, 2010 and June 15, 2012 organized by the United Nations under the authority of Ambassadors Kim Sook and John W. Ashe, co-chairs of the intergovernmental Preparatory Committee, and of the members of their Bureau. These meetings were the visible backbone of Rio+20. The process can be roughly divided into three stages.

The first stage was the preliminary intergovernmental discussions and preparations at the national, regional and global levels undertaken from May 2010 until the release of the Zero Draft of the Rio+20 Outcome Document, "The Future We Want," in January 2012. The second phase ran from the first round of informal negotiations that started in March 2012 and continued until the release of the co-chairs' streamlined text of "The Future We Want" in April 2012. The third and last phase of the formal preparatory process was dominated by long and tedious literal line-by-line negotiations and covered the second round of informal negotiations in April 9, 2012 up to the closing of the third preparatory committee meeting on June 15, 2012, when the negotiators were already in Rio de Janeiro.

### *Box 2.1* Outline of the Rio+20 formal preparatory and negotiations process

May 17–18, 2010: First preparatory committee meeting.

January 10–11, 2011: First intersessional meeting.

March 7–8, 2011: Second preparatory committee meeting.

November 1, 2011: Deadline for the submissions for the Zero Draft of "The Future We Want."

December 15–16, 2011: Second intersessional meeting.

January 10, 2012: Release of the Zero Draft of "The Future We Want."

March 19–23, 2012: First round of informal negotiations.

March 26–27, 2012: Third intersessional meeting.

April 9, 2012: Release of the co-chairs' streamlined text of "The Future We Want."

April 23, 2012–May 4, 2012: Second round of informal negotiations.

June 1–2, 2012: Third round of informal negotiations.

June 13–15, 2012: Third preparatory committee meeting.

June 16, 2012: Pre-conference informal consultations led by the host country, Brazil.

June 17, 2012: Release of pre-conference informal consultations draft of "The Future We Want."

June 19, 2012: Release and agreement by heads of delegations of the final text of "The Future We Want."

## *Lessons learned from the definition of the formal preparatory process*

When one is assessing why the third meeting of the preparatory committee did not manage, in its final days, to hand in an agreed negotiated document, it should not be overlooked that many of the factors hampering the effectiveness of Rio+20 preparations can be traced back to decisions unanimously made by government representatives during the winter of 2009 when they adopted the UNGA resolution for Rio+20.

In hindsight, the main lessons to be learned from the initial definition of the Rio+20 preparatory process are: (1) the duplication and – to a certain extent –

competition between the work of the United Nations Commission on Sustainable Development (CSD) and the Rio+20 process; (2) the establishment of a heavy intergovernmental negotiating machine; (3) the lack of enhanced involvement of Major Groups and other stakeholders who had not been involved in CSD meetings; (4) the limited number of days allocated by the UNGA to the preparations for the conference; (5) the limited financial resources dedicated to the informal-informals to enable more experts and stakeholders to participate; and (6) last, but not least, the definition of an ambiguous and broad mandate for the conference which reflected the underlying tension of giving a forward-looking vision to an event that had been initially called to commemorate achievements and assess the glaring gaps in the global implementation of Agenda 21.

The compromise made in 2009 was to set out a contained and limited process. The rationale was that, since the outcome of the conference was going to be a "focused political document," owing to strong lobbying from the United States, no extensive preparatory activities were required. In addition, the decision specified that neither the conference nor its preparations should undermine "the implementation of the multi-year program of work of the Commission on Sustainable Development and the overall expected appraisal in 2014 of the implementation of Agenda 21." This decision missed, from the outset, the opportunity to fully utilize the assets of the CSD for Rio+20 in order to build a comprehensive preparatory process. It also resulted in conflicting guidance towards the work of the CSD, since the objectives and the themes for Rio+20 were comparable, namely, assessing the progress in the implementation of Agenda 21 to date and reviewing the Institutional Framework for Sustainable Development (IFSD). Instead of adding in the weeks of the 2010 and 2011 CSD into the preparatory process and gaining an extra six weeks of already allocated time, representatives decided to establish a separate committee to carry out the preparations for Rio+20 while still trying to continue the CSD's multi-year programme. This decision further weakened the CSD, which already by 2009 was a body largely marginalized from major international processes.

The Rio+20 preparatory committee was open to all member states of the United Nations and members of the specialized agencies. This aimed at allowing the full inclusion of the State of Palestine (due to their membership of UNESCO) in its work. This composition – although it ensured universal representation and encouraged the participation of Major Groups and other stakeholders – was designed in a traditional way. Rio+20 missed the opportunity to "problem-solve" some of the challenges with a more interactive process before it moved to negotiations. The traditional approach also implied inherent limitations for the body to reach effective and bold decisions among 193 states.

At the helm of the preparatory committee, the UNGA established another heavy governing structure: a Bureau composed of ten members whose functions were not clearly defined. As the process evolved, this particular set-up placed constraints on the ability of the co-chairs of the Preparatory Committee to take difficult decisions concerning the conference, since these had to be made, in principle, in a consensual manner with the rest of the members of the Bureau. This reversed the approach taken within the CSD over the previous ten years whereby the chair of the Bureau did not have to reach consensus with the Bureau.

The UNGA also defined a limited calendar of preparatory activities consisting of two meetings in New York and one meeting in Rio de Janeiro before the beginning of the conference. For cost-cutting purposes, these were timed back to back with the eighteenth (May 2010) and nineteenth (March 2011) sessions of the CSD. The compromise reached initially allocated only eight days of work to the Rio+20 preparations – substantially less than the total of 25 working days allocated to the eighteenth and nineteenth sessions of the CSD. This cost-driven decision had a further disadvantage: rather than enticing a broad participation in the preparatory process of representatives from the social, development and economic ministries from capitals, it provided a significant incentive for maintaining the Rio+20 process only within the purview of national environment ministries and UN Permanent Representatives in New York, who had throughout the years grown accustomed to following the CSD sessions.

Concerned by this initial stringency, the first meeting of the preparatory committee held in May 2010 recommended the UNGA to organize three additional "informal intersessional meetings" to be timed according to the evolution of the process for a total duration of six working days. It also requested for the first time a substantive report from the Secretary-General on the themes of the conference (which we have reviewed in Chapter 1 of this book). By early 2012, once the Zero Draft of the Outcome Document, "The Future We Want," had been finalized, and following from pressure from the Group of 77, the Bureau conceded 17 additional working days for "informal-informal consultations."

In December 2009 the lack of a credible, agreed road map for Rio+20, with a realistic amount of time and resources, coupled with the different perceptions among developed and developing countries on whether the preparations for Rio+20 were "New York based" or "capital based," did not provide all negotiators with the same incentives to strike early agreements. As the months passed, the preparatory process lost significance, particularly as positions were repeated, new text introduced and brackets proliferated throughout the "streamlined compilation text." However, as the third round of informal consultations in New York drew to an end on June 2, 2012, negotiators shared the unspoken assumption that Rio+20 was simply too important to the United Nations and too politically visible to be allowed to fail.

Such unspoken convergence instilled among them the classic moral dilemma, with time constraints forcing negotiators to face the enormous task of striking last-minute agreements on unresolved issues in all the sections of the revised compilation Outcome Document. By that same token, there was also a general confidence that, at the end of the day, the Brazilian hosts were somehow going to save the show. Nevertheless, confronted by this dilemma, elements of the Document that had been previously agreed were actually revisited at the last minute; paragraphs on which there had been consensus were called into question (or "re-bracketed," to use international negotiation parlance), and issues that could have been agreed much earlier became victims of the desire to reopen dialogue rather than coming to closure.

Security, migration, oceans, gender, energy and the importance of the original Rio principle of common but differentiated responsibilities and respective capabilities were among the thematic areas affected by this malaise, which became so entrenched that one member state actually reserved its position on the entire section on "Means of Implementation."

To further complicate the problem, negotiators started introducing new ideas and initiatives for strategic purposes. One such example was the proposal for a Global Fund for Sustainable Development with an initial annual budget of US$30 billion (Watts, 2012).

At the same time, many well-intentioned Major Groups and other stakeholder representatives engaged with negotiators in the hope of securing a much sought-after reference to a specific issue in the ever-growing Compilation Document, arguing that without it the Rio+20 Outcome would remain "incomplete and unbalanced." What was interpreted by outside observers as irresponsible behaviour was in fact a way of signalling where key political divergences lay, in addition to being an encouragement for Brazil to produce, in due course, a text that would be acceptable to all.

## The political role of the UN Secretary-General

In analysing the Rio+20 Conference process and Outcome Document, it is necessary to fully understand the various roles played by the UN Secretary-General, Ban Ki-moon, in terms of his sustainable development initiatives, and by other key senior officers from UN agencies, which will be highlighted in the concluding chapter, Chapter 5, of the book.

From the perspective of Ban Ki-moon, Rio+20 can be regarded as a political success. But until the eleventh hour of the preparatory process, nobody within the secretariat was sure which way it would go. When Ban was elected as "the world's chief diplomat," in his very first address to the UNGA he described climate change as "the defining issue of our time," warned that its effects were being felt around the world and announced it as his top priority.

During that first term in office, the UN Secretary-General raised the attention and profile of climate change and convened a special session of the General Assembly to debate the issue at the highest political level (United Nations, 2008). Given the UN Secretary-General's stance and statements, there was global disappointment when the Fifteenth Conference of the Parties to the UN Framework Convention on Climate Change and the Kyoto Protocol

(UNFCCC COP15/MOP5) in 2009 in Copenhagen failed to reach consensus on a successor agreement to the Kyoto Protocol, or on the implementation policies and programmes to keep global temperatures from rising beyond the scientifically agreed "safe" limit of 2 degrees Celsius. Despite the UN Secretary-General's enthusiasm and efforts, the reality was that the multilateral system had proved unsuccessful in effectively addressing the urgency and gravity of climate change. This failure opened fissures in the fabric of multilateralism, which were causing people to question the continuing role and relevance of the United Nations and revisit the controversy on whether multilateralism could still deliver on global imperatives.

Before the start of his second term in late 2011, Ban stressed that "promoting sustainable development" was to sit at the top of his Five Year Action Agenda (United Nations, 2012e). Ban indicated that "[s]ustainable development will remain a key concern for the United Nations. Increasingly, it is needed to find sustainable solutions to guarantee development and economic growth around the world" (UN-NGLS, 2011). On the Secretary-General's list of priorities, sustainable development was followed by "preventing natural and man-made crises and disasters; making the world safer and more secure; helping countries and peoples in transition; and encouraging gender empowerment" (ibid.). Against this background, the inability of the multilateral system to reach agreement on an Outcome Document for sustainable development at Rio+20 would have signalled that, like the "top priority" of the UN Secretary-General's first term, the "top priority" of his second term also faced failure.

Adding to these policy and process concerns were two controversies and public relations debacles. In mid-2010 the UN Secretary-General was bitterly criticized in a 50-page memo authored by Inga-Britt Ahlenius, a Swedish auditor and outgoing UN Under-Secretary-General of the Office of Internal Oversight Services. Ahlenius accused Ban of poor leadership, allowing the UN Secretariat to fall into "a process of decay," as well as thwarting and circumventing her efforts on transparency and accountability (Lynch, 2010).

In September 2010, a mere two months after the publication of Ahlenius's damaging memo, there was a social gathering in Alpbach in Austria. Making a speech at the dinner was UN Under-Secretary-General and head of UN DESA Sha Zukang, who in May that year had been appointed secretary-general of the Rio+20 Conference and who had a reputation for being blunt and a maverick. In it he told the UN Secretary-General in front of the latter's guests (comprising senior UN staff) that although he had come to admire him, he, Sha, was in danger of being forced out of the UN system (The Australian, 2010; Daily Mail, 2010). In his comments, Sha Zukang was also severely critical of the United States, an important player to have at the table of the Rio+20 Conference of which he was secretary-general, and a key country as far as financial contributions for Rio+20 were concerned.

Despite this incident, there appeared to be no residual difficulties in the working relationship between Ban and Sha. They met for regular briefings and

Sha provided Ban with weekly updates on the progress of the preparations leading up to the conference. Interestingly, the speech was reported globally and did serve to turn a brighter and more colourful spotlight on Rio+20. Besides, coupled with a multiplicity of other issues, it resulted in increased personal involvement of the UN Secretary-General in the Rio+20 process and its outcome.

On a parallel track, the timing of the Rio+20 preparations was fraught with difficulty. An example of the complex circumstances was the clash with the Jubilee celebrations of the Queen of the United Kingdom and other Commonwealth realms. The result was a change of the date of the conference from early May to late June. This happened after the initially chosen date had already been publicized but it ensured that Commonwealth heads of state would be able to travel to Rio de Janeiro.

The failure to reach consensus in the preparatory meetings, and the early announcements during the preparatory process that the heads of countries important to the UN system – such as President Obama of the United States (because of upcoming elections), Chancellor Merkel of Germany and Prime Minister Cameron of the United Kingdom – would not be present at the conference, were circumstances that weighed heavily against a successful result. In addition to the timing and political problems, Rio+20 had also the challenge of overcoming a widespread image problem in the press, which predicted the conference was at worst doomed to be a failure and at best to perform below everybody's expectations (Black, 2012).

It was against this backdrop that the Rio+20 Conference took place, five months into the second term of Bank Ki-moon as Secretary-General of the United Nations. Given the challenges and priorities identified in his agenda for the second term, the Secretary-General had a lot riding on the outcome of the Rio+20 Conference.

As the date of the conference grew closer, Ban engaged directly and systematically with heads of state and government. He also held more and more regular meetings with the United Nations' senior leadership of the Rio+20 process and the Bureau in order to get updates on the organization, logistics, state of negotiations, dialogue, perceptions and positions of member states and other stakeholders. He also assigned various responsibilities to heads of agencies, such as the United Nations Development Programme (UNDP) and the United Nations Environment Programme (UNEP), in order to get the entire UN system to commit to delivering for Rio+20. As a leader, Ban was not only reactive but highly proactive in trying to put his own stamp on Rio+20, inspiring member states to deliver a strong development agenda and inserting innovations into the process, such as the establishment of the High-Level Panel on Global Sustainability (GSP), set up on August 9, 2010, which in fact diverted funding from the actual Rio+20 process.

Ban Ki-moon started the Rio+20 process in a precarious place, where each of those difficulties could potentially have dealt a crippling blow in defining his second term and derailing his agenda, particularly if another of his priority areas was seen to fail. In consequence, having the weight of the UN Secretary-General behind the preparatory process, themes and objectives of the Conference was critical. In perspective, Ban did rise to the occasion. In 2011 he reminded the UN membership that "[w]e need a revolution. Revolutionary thinking. Revolutionary action . . . [T]o make [sustainable development] happen we have to be prepared to make major changes – in our lifestyles, our economic models, our social organization, and our political life" (United Nations, 2011b). Strengthened by the results of Rio+20, the UN Secretary-General used his leadership to steward the broader UN system towards utilizing sustainable development as the overarching platform for building the Post-2015 Development Agenda, thus furthering from the institutional perspective the convergence between the planet-centred and the people-centred development agendas, making such achievement a cornerstone of his legacy to the Organization.

## Innovative partnerships supporting Rio+20

The key initiatives of the UN Secretary-General that underpin his vision of making sustainable development the core priority for the organization have been the Global Sustainability Panel (GSP), Sustainable Energy for All (SE4All), the Zero Hunger Challenge and the Oceans Compact. On the other hand, the Sustainable Development Solutions Network (SDSN) and the High-Level Panel of Eminent Persons on the Post-2015 Development Agenda were initiatives launched after Rio+20 and in support of the post-2015 agenda, and can be interpreted as in favour or not of supporting the Rio+20 follow-up process.

### *The Global Sustainability Panel*

The role of the GSP was to bring together important world figures "to formulate a new blueprint for a sustainable future on a planet under increasing stress resulting from human activities" and "to explore approaches for building a low-carbon, green and resilient economy that can eradicate poverty and ensure a dignified life for all." It was intended that the GSP, officially set up in August 2010, would also serve to champion the cause of sustainable development, broaden the dialogue and thought processes, and then issue a final report in early 2012 that would feed into the intergovernmental preparatory process for Rio+20, as well as into the annual meetings of the United Nations Framework Convention on Climate Change (UNFCCC). The GSP comprised 21 high-level members, including five heads of state or government (United Nations, 2011a).

The inspiration for the panel drew from many intellectual and political sources. Its essence was inevitably reminiscent of the original World Commission on Environment and Development, which gave birth to the iconic Brundtland Report. It also drew from contemporary examples – such as the 2006 High-Level Panel on UN System-Wide Coherence – and recommendations – such as the 2008 civil society recommendation for a new World Commission on Sustainable Development made in the Donostia Declaration in November 2008. This declaration stated that "to mobilize new thinking the UN Secretary-General should explore the possibility of a World Commission on the Green New Deal which would report to the Summit process in 2011 and 2012" (Stakeholder Forum, 2008). The idea of a commission was also embraced during 2009 by a number of key advocates. Possibly the most significant endorsement it received was from the Swedish-sponsored Commission on Climate Change and Development Report, entitled "Closing the gaps," whose lead author was Lloyd Timberlake, one of the authors of the 1987 Brundtland Commission report. This report

> recognize[d] the leadership of the SG [Secretary-General] and urged [him] to convene, in cooperation with international financial institutions, an independent high-level task force to articulate a vision for development that achieves the multiple goals of mitigation, adaptation, and meeting human needs. High on its agenda should be the connections between the different global crises and the adequacy of global public policy and global governance in dealing with them simultaneously.
>
> (Commission on Climate Change and Development, 2009)

These recommendations were taken up by the UN Secretary-General, who in a rather vague way in September 2009 announced his intention to set up a panel on sustainable development during his closing remarks at the climate change event in the UNGA in New York.

Other key issues to solve were the appointment of the head of the secretariat of the panel and the decision on whether to place this head within the secretariat for Rio+20. If this did not happen, then what relationship would the Panel Secretariat have with the Rio+20 Secretariat?

Janos Pasztor, who had been working as director of the UN Secretary-General's Climate Change Support Team, preceding the Copenhagen Climate Change Conference, COP 15 (2009), was tasked with coming up with a concept for this new initiative on sustainable development. He started the process in 2010 but was hampered by a number of problems. The first problem was identifying what the panel should address. To begin with, the whole panel concept had more of a climate change focus. It became broader as the realization grew that climate change cannot be solved by itself, and in view of the Cancún Agreements, the scope of the panel needed to be broader. Even the final terms of reference of the GSP had more of a focus on climate change than the final report, which really has very little. The importance of the economic dimension of sustainability became increasingly clear with the 2008 financial crisis and its aftermath. But, unfortunately, the GSP did not have enough economic expertise to make that element stronger and more specific.

It took time to put the terms of reference and the panel members together, and its announcement coincided with the opening of the Rio+20 debate. This created unfounded fears among several member states that the UN Secretary-General was trying to bypass the official negotiation process and come up with a separate blueprint for Rio+20. The Panel members and the staff made remarkable efforts to allay these fears and ensure transparency. This also led to the proactive involvement of the so-called panellists' Sherpas in the drafting and eventually the negotiating of some of its main recommendations.

There was also considerable discussion about who would chair the panel. There were several alternatives discussed for providing the necessary leadership to the GSP. Eventually, the President of South Africa, Jacob Zuma, whose country had been the last to host a summit on sustainable development, and the President of Finland, Tarja Halonen, whose country had become a champion of sustainability and effective international cooperation, agreed to lead the panel. The panel also had some clear heavyweights from all sectors, such as Gro Harlem Brundtland; James Balsillie, who was then head of Research in Motion (better known as RIM, the creators and manufacturers of BlackBerry mobile phones); Connie Hedegaard, the European Commissioner for Climate Action; Susan Rice, US ambassador to the United Nations; Kevin Rudd, Minister of Foreign Affairs of Australia; Jairam Ramesh, Minister of the Environment of India; and Julia Carabias, a highly respected academic and environmentalist who had also served as Minister of the Environment of Mexico. Izabella Teixeira, Minister of the Environment of Brazil, was added to the GSP membership several months after the panel's launch, showing the growing importance of Rio+20 in the thinking of the GSP and the increasing acceptance of the former in its work.

The panel worked for 18 months in parallel to the Rio+20 process, and on January 30, 2012, in Addis Ababa, Ethiopia, it launched the report *Resilient People, Resilient Planet: A Future Worth Choosing* (United Nations, 2012c). The vision for the forthcoming decades laid out by the report was "to eradicate poverty, reduce inequality and make growth inclusive, and production and consumption more sustainable, while combating climate change and respecting a range of other planetary boundaries." This vision – influenced by the thinking of two of its leading members, Kevin Rudd and Jairam Ramesh – is grounded on two simple questions: if sustainable development provides the best opportunity for people to thrive and to choose their future, then why are people not choosing it? Why has sustainable development not become a reality in every society and in every individual's life?

The answer provided by the panel to this question pointed to the fact that, almost three decades after the Brundtland Report, a new approach to sustainable development was needed. This approach is grounded in opportunities and choice (allow the people the opportunity to choose) as well as in institutions, understood not simply as organizations but also as norms, rules and values.

The 1987 Brundtland Report's vision was that sustainable development could be achieved by an "integrated policy framework" embracing all three pillars (economic, social and environmental). The world did not completely fail in this regard: it did develop and agree Agenda 21, the ultimate global sustainable development framework. Since then, the world has gained a deeper understanding of the interconnected challenges we face and the dynamics for the decades to come. The truth is, GSP members said, that sustainable development is fundamentally a question of people's opportunities and rights to influence their future, claim their rights and voice their concerns vis-à-vis their governments and the corporations breaching their compact with them as custodians of a sustainable future for all. Real opportunities for choosing sustainable lifestyles could only be achieved by "a new political economy." This meant

> radically improving the interface between environmental science and policy; recognizing that in certain environmental domains, such as climate change, there is "market failure," which requires both regulation and what the economists would recognize as the pricing of "environmental externalities," while making explicit the economic, social and environmental costs of action and inaction; recognizing the importance of innovation, new technologies, international cooperation and investments responding to these problems and generating further prosperity; recognizing that an approach should be agreed to quantify the economic cost of sustained social exclusion – for example, the cost of excluding women from the workforce; recognizing that private markets alone may be incapable of generating at the scale necessary to bring about a proper response to the food security crisis; and requiring international agencies, national Governments and private corporations to report on their annual sustainable development performance against agreed sustainability measures.
>
> (United Nations, 2012c)

Against this vision, the GSP's report was broken down into six broad sections which followed the scope and main themes for Rio+20: vision; progress towards sustainable development; empowering people to make sustainable choices; working towards a sustainable economy; strengthening institutional governance; and a call for action. The report's findings and recommendations fell into three areas (United Nations, 2012c):

1 Recommitment to previous agreements. These included the Millennium Development Goals (MDGs), human rights agreements, gender commitments, regional fisheries and marine commitments.
2 Support for existing discourse around Rio+20. This included the strengthening of UNEP; the creation of a Sustainable Development Council; Sustainable Development Goals (SDGs); a framework for corporate reporting; the phasing out of fossil fuel subsidies; the Earth Charter Initiative;

universal access to quality and affordable family planning, and other sexual and reproductive rights and health services; a chief scientist for the UN Secretary-General Secretariat; and a global sustainable development outlook report.

3 New ideas in the panel's report, the most notable being the section devoted to the use of sustainable development criteria in the management, rating, risk assessment, and regulation of and investment by decisions of private and sovereign capital to enable the transition to sustainable development. These measures were receptive to the major failures and deficiencies of credit rating agencies – themselves publicly owned companies – during the financial and economic crisis of 2008. In this regard, the panel spoke beyond the expected audience for Rio+20 and provided recommendations addressed to the boards of sovereign wealth funds, national and international public pension funds, and government agencies responsible for regulating the stock markets.

The main problem with the panel's 49 recommendations was that they were presented far too late in the Rio+20 process. An emblematic illustration is the development within the GSP framework of the idea on Sustainable Development Goals. The report was actually published at the end of the month in which the Zero Draft of the Rio+20 Outcome Document was released to the public and a deluge of inputs to it was already flooding the UN DESA website designated for the purpose.

At the time of the launch of the report, the first negotiating session on the Zero Draft had already taken place. This timing meant that the GSP report missed the critical window for influencing the Rio+20 preparation process and Outcome Document. To be more effective, the report should have come out in the summer of 2011, which would have enabled governments and the UN system to engage in more informed deliberations, and could even have provided language for negotiators. By the time the GSP's report came out, governments had already outlined and prioritized their ideas, which left minimal room for new concepts to be integrated.

Nevertheless, on a brighter note the GSP's report did give legitimacy and support to some of the ideas that were being floated but for which core support had not yet coalesced, such as the SDGs and SE4All. The GSP's endorsements inevitably helped some of these ideas gain acceptance and make their way into the final text of "The Future We Want."

## *Sustainable Energy for All (SE4All)*

In September 2011 the UN Secretary-General's initiative Sustainable Energy for All (SE4All) was launched (www.sustainableenergyforall.org). As a complement to the Initiative, the UNGA proclaimed 2012 as the "Year of Sustainable Energy for All." The Year was intended to "mobilize actions from governments, business and civil society to provide a vital advocacy platform for raising awareness of the challenge" relating to energy. This initiative was a concretization of one of the themes that had attracted significant attention in the 2009 Financing for Development Review Conference in Doha, in the 2010 MDG review and in several pre-Rio+20 dialogues, when every political grouping within the UN system identified energy as a priority and one of the major drivers in achieving sustainable development.

The SE4All Initiative was first established with a High-Level Panel and a supporting Technical Committee, which developed an 11-point "global action agenda" with an "implementation road map." The strategic approach included metrics and was intended to provide "a framework for identifying high impact opportunities, . . . a way to organize multistakeholder actions and . . . a tangible entry point for stakeholders." The initiative had three interlinked objectives to be achieved by 2030:

1 ensuring universal access to modern energy services;
2 doubling the share of renewable energy in the global energy mix;
3 doubling the global rate of improvement in energy efficiency.

Kandeh Yumkella, the indefatigable former director-general of the United Nations Industrial Development Organization (UNIDO), led the work on the initiative and later took on that assignment on a full-time basis as head of an especially dedicated Secretariat. From the private sector, and appointed as co-chair of the initiative, was Charles "Chad" Holliday, chair of the Bank of America.

Given the data on energy poverty globally, an initiative on energy was considered not only necessary but urgent. Those statistics included the facts that one in five people across the world have no access to electricity, 3 billion people are dependent on coal, wood, charcoal and animal waste for cooking and heating, and energy as a dominant contributor to climate change causes 60 per cent of greenhouse gases. The UN system rapidly bought into the initiative, with more than 20 agencies joining in the work towards its successful implementation.

SE4All received a boost on January 30, 2012, when the GSP's report endorsed the initiative in its paragraph 49: "the Panel endorses the UN Secretary-General's 'Sustainable Energy for All' initiative, which should be implemented without delay" (United Nations, 2012c). The reinforcing goals of the initiative were also identified by the GSP and by the members of the "group of friends" of sustainable energy (a grouping addressed later in this chapter) to be early candidates under the initiative for establishing Sustainable Development Goals.

Despite this unequivocal endorsement from the GSP, in the run-up to the Rio+20 Conference there was reticence in the reaction by member states to the initiative and it was not endorsed or adopted at any of the regional or preparatory meetings for Rio+20 (except for the one in Barbados), or in the negotiations prior to the conference. In this early period it appeared as if SE4All would fall irretrievably into the chasm between developed and developing countries on energy issues, as well as between the proponents of renewable energy and those who felt that oil still had a bright future.

Prior to Small Island Developing States' launch of SE4All in May 2012 in Barbados, the panorama was marked by the absence of specific commitments by member states to the initiative, its lukewarm welcome at Rio+20 negotiations and the failure to definitively commit or agree to its objectives in all the regional and preparatory meetings. At the SIDS launch the prime minister of Barbados, Freundel Stuart, in clear and unequivocal language, broke the pattern of absence of support.

In framing his perception of the importance of SE4All, the UN Secretary-General commented at Rio+20 that

> more than one hundred commitments and actions have been already mobilized in support of Sustainable Energy for All, demonstrating powerful early momentum from governments, private sector companies and civil society organizations to achieve Sustainable Energy for All by 2030 . . . Achieving sustainable energy for all is not only possible, but also necessary. It is the golden thread that connects development, social inclusion and environmental protection . . . This initiative is already mobilizing significant action from all sectors of society. Working together, we can provide solutions that drive economic growth, expand equity and reduce the risks of climate change.
>
> (UN News Centre, 2012a)

The United Nations and the World Bank signalled their intention to work together on the attainment of SE4All, and at the Spring Meetings of the World Bank on April 19, 2013 the leaders of the two organizations, UN Secretary-General Ban Ki-moon and World Bank president Kim, announced an Advisory Board of 42 members drawn from business, finance, government and stakeholders from varied disciplines and geopolitical regions. The announcement indicated that the Advisory Board "will provide strategic guidance and serve as global ambassadors for the Initiative conducting high level advocacy for action on energy and mobilization of stakeholders on behalf of SE4All." Noting that some 70 countries had opted into the initiative since its launch, the UN Secretary-General described the continuing work as "a global public–private sector partnership."

When referring to energy, the UN Secretary-General described sustainable energy as a "golden thread." As time progresses, it will be interesting to observe how effectively and beneficially this "golden thread" is used by the SE4All Secretariat, member states, the private sector and private citizens to bind together the three dimensions of sustainable development and poverty eradication.

Support for SE4All was not uniform: as Rio+20 started, 25 non-governmental organizations called for support for "a more ambitious, accountable and people-driven Sustainable Energy for All initiative." Led by the World Future Council, Greenpeace International and Friends of the Earth, they said:

> We are concerned that while corporate interests from the fossil fuel, finance and energy industries have been invited by the UN Secretary-General to his High-Level Group most national governments and civil society voices have not. Civil society and the energy poor have generally been accorded a limited role throughout SE4All. Any national strategy must lay out clearly defined accountability and monitoring mechanisms to enable citizens to hold their government and any private sector concessionaires accountable in real time for progress on commitments and their impacts. We call on donor and private sector actors to commit to funding such a process of civil society engagement, accountability and monitoring in SE4All countries to ensure the effectiveness of national roll-out.
>
> (World Future Council, 2012)

Despite the slow start in terms of support, "The Future We Want" eventually contained language on energy, committing to principles and activities analogous to those of the SE4All initiative, though without specifically saying that it was committing to SE4All. Paragraphs 125–129 of "The Future We Want" speak to the importance of access to sustainable modern energy resources, increased energy efficiency and the availability of renewable energy technologies. The Outcome Document of Rio+20 also speaks directly to the Sustainable Energy for All initiative, and expressed "determination to act to make sustainable energy for all a reality and, through this, help to eradicate poverty and lead to sustainable development and global prosperity." SE4All also attracted the second largest number of voluntary commitments of the Rio+20 Conference and the largest number of financial commitments made for any sector or thematic area. Commitments were made by all stakeholders, including development banks, national governments, the private sector, academia and the NGO community, on SE4All ranged from policy to financing projects across the three objectives of the initiative. The importance of energy to the development agenda and the potential of SE4All is reflected in the actions and activities that have continued to take place since Rio+20 in the context of the preparations for the Post-2015 Development Agenda. Energy was included among the 11 issues for which the UN system organized global thematic and regional consultations.

## *The Zero Hunger Challenge*

The Secretary-General raised the profile of the interlinkages between food security and sustainable development through the Zero Hunger Challenge (United Nations, 2012f). This is an important initiative in a world where

1 billion people are still hungry at the end of every day and in which it is expected that the global population will grow to 9 billion by 2050 (Parker, 2011). The central thesis of the Challenge is that "hunger can be eliminated in our lifetime" by utilizing "comprehensive efforts to ensure that every man, woman and child enjoy their Right to Adequate Food; women are empowered; priority is given to family farming; and food systems everywhere are sustainable and resilient." The Zero Hunger Challenge entails the following objectives:

- 100 per cent access to adequate food all year round;
- zero stunted children aged less than 2 years;
- making all food systems sustainable;
- a 100 per cent increase in smallholder productivity and income;
- zero loss or waste of food.

The Challenge supports "investments in agriculture, rural development, decent work, social protection and equality of opportunity." It asserts that a world without hunger "will contribute to peace and stability and to the reduction of poverty [and] to better nutrition for all – especially women from the beginning of pregnancy and children under the age of two" (United Nations, 2012f).

### *The Oceans Compact*

The UN Secretary-General's leadership since the conference can also be found in the initiatives that he has continued to pilot and on which he is seeking to build momentum. Oceans were one of the major topics at Rio+20, where the new expression "blue economy" was introduced into the lexicon of sustainable development. On January 25, 2012 the UN Secretary-General released his "Five-Year Action Agenda," in which he said on oceans: "Agree to a compact on oceans that will address overfishing and pollution by improving the governance of oceans and coastal habitats and by developing an institutional and legal framework for the protection of ocean biodiversity" (United Nations, 2012e).

In January 2012 he also surprisingly announced a five-person expert group to develop this idea. In section V of "The Future We Want," the words "commit" and "commitment" were used more frequently in regard to oceans than in connection with any other issue. Two months after the conclusion of the Rio+20 Conference, UN Secretary-General Ban Ki-moon launched his Oceans Compact in his native Korea (UN News Centre, 2012b; United Nations, 2012d). On the occasion of the launch, he observed:

> The seas and oceans host some of the most vulnerable and important ecosystems on Earth, but the diversity of life they host is under ever-increasing strain . . . What we need is to create new momentum for ocean sustainability . . . The Oceans Compact sets out a strategic vision for the

> UN System to deliver more coherently and effectively on its oceans-related initiatives.
>
> (ibid.)

Ban saw the Compact as "a platform to help countries protect the ocean's natural resources, restore their full food production to help peoples whose livelihoods depend on the sea, and increase awareness and knowledge about the management of the oceans." To achieve the objectives of the Compact, the UN Secretary-General proposed a results-oriented Action Plan along with the creation of an Ocean Advisory Group made up of high-level policy-makers, scientists and experts, as well as representatives of the private sector and other stakeholders.

## The informal process in New York and its importance for consensus-building

In parallel with the formal process for Rio+20 there were a number of informal yet highly important initiatives that convened in New York with the aim of influencing the preparatory process of Rio+20. These exercises were for the most part held as issue-based, cross-regional "groups of friends." They served specific purposes: they provided a space for delegations to share information; fostered understanding on specific themes; allowed engaged ambassadors to advocate, and play a leading role; provided a space for Major Groups and other stakeholders to present expert input and advice; and ultimately, in several successful cases, facilitated cross-regional consensus. A limiting factor was that these were not open processes, so those not invited had no chance to affect the discussion.

Looking back, it is possible to appreciate that most of the issues included in the Rio+20 Outcome Document in section III on the green economy, section IV on the Institutional Framework and section V on Thematic Areas were initially developed in the context of these informal processes. Although informal constituencies have always played a role in the work of the United Nations, Rio+20 constitutes an exception because of the diversity of themes, the level of political engagement and the richness of independent advice used to support each of these ad hoc processes. In a way, this proliferation is an illustration of the difficulties for advancing new sustainable development issues within the formal UN intergovernmental process and the growing complications faced by delegations to build agreements and allow truly interactive dialogue among ambassadors and senior negotiators in plenary settings, the format of which has become increasingly ritualized and rigid. Finally, the emergence of these informal processes also signals the interest of many countries in working in cross-regional settings beyond the confining and often adversarial boundaries of established constituencies.

As a testament to their effectiveness, the thematic areas and cross-sectoral issues included in "The Future We Want" that were supported by specific groups of friends were poverty eradication and the MDGs; food security, nutrition and sustainable agriculture; water and sanitation; energy; sustainable tourism; sustainable cities and human settlements; oceans and seas; disaster risk reduction, gender equality and women's empowerment; and mountains. In addition to these, there were other regular informal consultations for discussing broader issues such as the Rio+20 preparations, the green economy, the Institutional Framework for Sustainable Development and the Sustainable Development Goals.

The informal preparatory process in New York also allowed for other, more timely informal discussions to build consensus around the inclusion in section V of "The Future We Want" of some of the themes of the agenda of the failed nineteenth session of the Commission on Sustainable Development, namely transport, chemicals and waste, mining, and sustainable consumption and production. This chapter will summarize the main scope of, and leading countries behind, these informal preparatory activities.

### *Informal contact group on the MDGs*

The informal dialogue on the poverty eradication agenda, the Millennium Development Goals (MDGs), kicked off in New York in December 2011 on the initiative of Japan and was followed up through the work of an informal contact group. This group was described as a forum for informal policy dialogue on the development agenda beyond 2015. Its work gathered participants from about 20 countries, as well as major international organizations, foundations, research institutions and NGOs. The informal contact group held four meetings in Mexico City, New York and Tokyo between December 2011 and September 2012.

Upon the conclusion of the first phase of its work, Japan published a summary of the contact group in which it stated that the fundamental priority of the poverty eradication agenda remained the accelerated implementation of the MDGs. It also stated that the collective aspiration of the participants of the group was to build on the momentum given to sustainability by Rio+20 towards a new development agenda; highlighted a strong argument that any new development framework "should be about people and the planet," and said that "3Cs (Coherence, Coordination, Convergence) are of fundamental importance." The post-2015 development framework should, according to its organizers, "maintain the strengths of the current MDGs framework, with a focus on poverty reduction as well as social and human development." Finally, the preliminary conclusions called for a single framework that incorporates sustainability, while maintaining simple, clear, easy-to-understand and measurable goals (Japan, 2012).

One of the impacts of this informal contact group on the Rio+20 process was the increased awareness of the risk of fragmentation of the global development agenda, particularly if Sustainable Development Goals were agreed ahead of the completion of the MDGs. The contact group provided a space for the participation of representatives of the so-called development community, especially international development ministries, international financial institutions and other actors from developing countries engaged in the coordination and follow-up of the MDGs, many of which had not proactively engaged in the early phases of the Rio+20 preparatory process, since it was perceived by them as an "environmental" process.

### *Friends of food and nutrition security*

The group of friends of food and nutrition security was launched in 2011 upon the initiative of the Netherlands' ambassador to the United Nations at the time, Herman Schaper. Originally it attracted more than 23 countries as well as the major UN entities working on issues related to food security. Its purpose was to inform, essentially through expert panels, the Rio+20 negotiations. In addition to regular informal discussions involving ambassadors and policy experts, it organized outreach events in New York during the second round of consultations in late April 2012.

### *Friends of water*

Building on the long-standing engagement at UN level of Tajikistan on the issues related to water, the ambassadors of Hungary, Thailand and Finland joined efforts in early February 2012 to launch a group of friends to discuss water-related issues in the context of Rio+20. The idea was initiated by this group of countries with the goal of "making a positive impact and supporting pragmatic, result and solution-oriented input to the Rio+20 Outcome Document and conference preparations."

The main outcomes of the group were a non-paper on the lessons learned from five expert panels organized by the key countries involved in steering the group, in addition to language proposals drafted for the Rio+20 Outcome Document. This informal platform was an effective forum that reached out to the various national perspectives, including at ministerial level, on issues related to water management and cooperation. It also provided a space for interaction between countries, water operators and civil society organizations. Finally, it was helpful in introducing to the Rio+20 process the growing scientific evidence on the nexus between fresh water, food security and energy.

### *Friends of sustainable energy for all*

On March 2011, Ambassadors Morten Wetland of Norway and Carsten Staur of Denmark, joined by Ambassador Abdullah Hussain Haroon of Pakistan, seized the opportunity ahead of Rio+20 to launch an ambitious effort to provide in

New York a "home" for energy-consideration and informal policy discussions among countries.

The stated goal for the group of friends of sustainable energy for all was to serve as "a sounding board for ideas developed by different actors within the UN system," particularly from the Department of Economic and Social Affairs (UN DESA), the United Nations Industrial Development Organization (UNIDO), the former UN Secretary-General's Advisory Group on Energy and Climate Change and UN Energy, among others.

The group reached out initially to close to 30 key countries and gradually expanded the scope of its participation. It was also able, through regular expert and political engagement, to support the launching in Oslo in October 2011 of the UN Secretary-General's initiative Sustainable Energy for All. From March 2012 onwards, this group of friends concentrated its activities on developing ideas for areas to improve the Zero Draft of "The Future We Want" and remained engaged, albeit with less dynamism, in the final course of the negotiations.

The contribution of the group was nevertheless fundamental for evolving, in the context of the post-2008 financial and economic crisis, the discussion on sustainable energy as a key driver to the achievement of the MDGs. This built on the progress achieved in that regard in 2008 during the Doha Review Conference and on 2010 in the MDG Review Summit. The work of this group highlighted, in "The Future We Want," a shared global commitment "to act to make sustainable energy for all a reality, and through this, help to eradicate poverty and lead to sustainable development and global prosperity" (United Nations, 2012b).

The group also established itself beyond Rio+20 as a key New York-based informal platform for engaging broader political support towards a commitment, within the UN development agenda beyond 2015, for setting specific goals on universal access to modern energy services, increasing energy efficiency and increasing the use of renewable sources of energy, building on the commitment reached in Rio+20. Its work and activities have evolved largely evolved around the progress of the UN Secretary-General's initiative.

### Friends of sustainable tourism

Sustainable tourism had some recognition in Rio+20, thanks to the ownership of this by a group of ambassadors belonging to the Central American Integration System, a regional grouping integrated by Belize, Costa Rica, the Dominican Republic, El Salvador, Guatemala, Honduras, Nicaragua and Panama. Their efforts were also supported by Morocco, which has been a long-standing supporter of highlighting, through the work of the UNGA, "the potential of ecotourism to reduce poverty by improving individual livelihoods in local communities and to generate resources for community development projects" (United Nations General Assembly, 2013).

### *Friends of sustainable cities*

The ambassador of Singapore, Albert Chua, and Sweden's ambassador for sustainable development, Staffan Tillander, launched in November 2011 – in time for the second intersessional meeting of the Rio+20 preparatory process – a group of friends of sustainable cities which played a critical role in raising awareness of the need to include the impact of urbanization and the role of sub-national and local authorities in the sustainable development agenda.

Owing to the support given by the United Nations Human Settlements Programme (UN-HABITAT), and thanks to its openness in accepting support from the Local Authorities Major Group, the group contributed to shaping, in "The Future We Want," an innovative way of looking at the opportunities – rather than just the challenges – that urbanization provides for achieving sustainable development. The key message that the conveners of the group put forward by April 2012, and that eventually made its way into framing the section on cities and human settlements of "The Future We Want," was that "sustainable cities can be and sometimes already are environmentally sustainable, socially responsible and economically productive" (UN-HABITAT, 2012).

This group of friends, which held regular expert meetings between November 2011 and late April 2012, contributed to Rio+20 through highlighting key issues that form part of a new urban agenda. Some of the issues raised were the need to elaborate national urban development policies; the need to increase an integrated and multilevel governance approach to urban planning and development in order to empower sub-national and local authorities, including through financial means; the importance of improving living conditions in cities by upgrading basic services such as those concerned with health, nutrition, safe potable water, sanitation and waste management; and the need to implement land policy development and regulatory and procedural reform programmes, in addition to advancing the right to adequate housing, promoting citizen participation and establishing comprehensive risk management strategies to ensure greater resilience to natural disasters.

### *Friends of oceans and seas*

The group of friends of the oceans was an initiative launched in July 2011 by Ambassadors Gary Quinlan of Australia and Joseph Goddard of Barbados. In its initial stages the group involved more than 20 ambassadors and counted support from the Pew Environment Group. Its initial aim was "to provide an informal platform for dialogue on blue economy issues in the lead-up to Rio+20," mainly through sharing national policies, expertise and information. One of its purposes was to highlight the key concern among Pacific island countries and other Small Island Developing States, which felt that the profile of the oceans agenda was not receiving sufficient attention in the Rio+20 preparatory process. In this regard, the group of friends built upon the work already undertaken by Australia during the 2010 and 2011 sessions of the General Assembly, based on the dedicated efforts of Katy Lin, one of the most committed Australian negotiators.

This process made it possible to table, with broad international support, two landmark UNGA resolutions entitled "Protection of Coral Reefs for Sustainable Livelihoods and Development" (UNGA 65/150 and UNGA 66/194; United Nations General Assembly, 2011, 2012). These resolutions paved the way to address, from the perspective of sustainable development, issues connected with oceans, corals and related ecosystems in Rio+20. They also contributed to reinforcing a sense of urgency among member states "to identify relevant measures or tools for the protection of coral reefs within their national jurisdiction as an urgent sustainable development priority."

In this regard, Australia and Barbados, the conveners of the group of friends, expressed the view that, while recognized in Agenda 21 and the World Summit on Sustainable Development Plan of Implementation, the nature and scale of the challenges facing the oceans, and the implications for sustainable development, made oceans and seas a key emerging issue and a priority future challenge. A particularly difficult topic for which the group of friends attempted to provide a space for discussion was that of the sustainable use and conservation of marine living resources of the high seas, which has remained one of the most glaring unimplemented commitments contained in chapter 17 of Agenda 21.

In its work, this group of friends engaged countries, oceans and fisheries experts from UN agencies, NGOs and think tanks. It also provided a first informal platform to air concerns of some delegations in New York, and to complementary platforms launched in the wake of Rio+20 such as the World Bank-led initiative Global Partnership for Oceans and the UN Secretary-General's Ocean's Compact.

The section on oceans and seas of "The Future We Want" was among the last issues to be resolved during the final stages of the Rio+20 negotiations, owing to the fact that the "problems of ocean space are closely interrelated and need to be considered as a whole" (UNGA 65/150 and UNGA 66/194), particularly in the context of the follow-up processes to the United Nations Convention on the Law of the Sea, which traditionally has had a pronounced juridical approach. The work of this group of friends nevertheless contributed to raising the profile of such issues from the sustainable development perspective and helped in achieving a commitment to take, before the end of 2015, "a decision on the development of an international instrument under the Convention on the Law of the Sea" for the conservation and sustainable use of marine biodiversity beyond areas of national jurisdiction. This work was bolstered by two initiatives (discussed elsewhere in the book): the marine and fisheries sustainability initiative of His Royal Highness Prince Charles of the United Kingdom, and the oceans and seas initiative and workshop hosted by His Serene Highness Prince Albert II of Monaco.

### *Friends of disaster risk reduction and resiliency to natural disasters*

The group of friends concerned with reducing the risk of disasters was a joint cross-regional initiative convened by Ambassadors Desra Percaya of Indonesia, Morten Wetland of Norway, Enrique Román Morey of Peru and Gary Quinlan of Australia. Japan was also active in it as part of its bid to host, in early 2015, the Third World Conference on Disaster Risk Reduction.

The objective of the group was twofold: (1) to discuss, in the context of sustainable development, the critical elements of the disaster risk reduction agenda; and (2) to identify and support opportunities, through joint statements and side events, to raise awareness of disaster risk reduction throughout the Rio+20 process.

The political outreach effort in New York was reinforced by the Geneva-based support group of the United Nations International Strategy for Disaster Risk Reduction (UNISDR). Both informal platforms served as tools to help countries come together to advocate collectively for a better understanding of the importance of having disaster risk reduction strategies in both global and national developmental plans. With this in mind, from late 2011 the group of friends for disaster risk reduction convened informal meetings on a regular basis. By reaching out early to member states, this group of friends became critical in building the consensus reached in "The Future We Want" on disaster risk reduction as a fundamental tool to retain the gains resulting from implementation of the MDGs and to achieve resilience and sustainable development. Such mobilization by like-minded delegations was also critical in ensuring the incorporation of disaster risk reduction as a necessary component of the Post-2015 Development Agenda.

### *Friends of the mountains*

Prior to Rio+20 the group of friends of mountains was revitalized, owing to the initiative of the ambassadors of Nepal, Switzerland, Peru and Austria. Their initiative aimed at highlighting in the Outcome Document of Rio+20 the contribution of mountain countries and mountains regions in addressing common problems. Two specific issues that this group served to highlight were the vulnerabilities of mountain countries with regard to climate change and their potential to contribute to the sustainable development agenda. Their efforts as a group were articulated through the drafting and subsequent sponsoring of UNGA Resolution 66/205 on Sustainable Mountain Development (adopted in December 2011), and the Kathmandu Call for Action, adopted at the International Conference of Mountain Countries on Climate Change held on April 5 and 6, 2012 in Kathmandu, Nepal.

### *Friends of gender equality and women's empowerment*

Gender equality and women's empowerment were well integrated into "The Future We Want," owing to concentrated efforts by key delegations and civil society gender advocates, as well as by support from the UN system. Notably, the delegation of Norway via Marianne Loe, senior negotiator for Rio+20, and Tine Mørch Smith, deputy ambassador in New York, carried recommendations on gender equality and women's empowerment drawn directly from the Global Sustainability Report (GSP recommendations 2 and 3), which became the anchor for the gender section of the Rio+20 outcome.

The Icelandic delegation also played a pivotal role throughout the process, ensuring that gender was identified as a cross-cutting issue and integrated into multiple sections (e.g. energy, disaster risk reduction, education), as well as also advancing the issue of women's equal participation in decision-making and management at all levels.

A particular improvement in sustainable development and gender equality was the text (introduced by Iceland) on temporary measures for achieving gender parity contained in paragraph 237 of "The Future We Want." Finally, the Women's Major Group actively advocated and provided support to delegations, as did UN Women, which held a series of informal meetings with member states during the preparatory process.

### *Friends of Rio+20*

The group of friends of Rio+20 was the conference's first informal group. Created in early 2010 upon the initiative of two committed negotiators, Claire Thuaudet from France and Randhir Jaiswal from India, and being a cross-regional group of UN delegates, it allowed informal discussion of the preparations for the Rio+20 Conference. The key motivation of that group, which gathered only at expert level and on a regular basis in 2010 and 2011, was the common interest shared by all of its members in making Rio+20 a success. It helped in sharing ideas, information and impressions on the process and the goals of the conference. It was also particularly useful in promoting a better understanding of the objectives, the main themes and the organizational modalities of the conference, particularly during the negotiations in 2010 and 2011 in the UNGA on the guiding resolutions for the conference.

### *Friends of the UNGA thematic debate on the green economy*

Discussions in New York on the green economy kicked off before any decision was made to convene Rio+20. However, it was not until January 2011 that a first thematic group of friends was created upon the initiative of Ambassador Paul Seger from Switzerland and involving a group of ambassadors from key countries such as Egypt, Gabon, Grenada, Hungary, Indonesia, Mexico, the

Republic of Korea, Romania and Sweden. The group's main objective was to build – both political and substantive – momentum towards an informal thematic debate of the UNGA held in June 2011 under the patronage of Joseph Deiss, president of the General Assembly (Deiss, 2011).

The Swiss-led initiative benefited in addition from a substantial intellectual backing by the World Resources Institute (WRI). This collaboration allowed the group of friends to organize, between February and May, three informal workshops and one ambassadorial retreat. These workshops provided representatives in New York with a first platform to have an informed and experience-based dialogue on a concept that had been until then, even despite its inclusion in the outcome of the Conference on the World Financial and Economic Crisis, completely unfamiliar to the United Nations.

This group of friends and the informal thematic debate of the UNGA helped build a better understanding on the principles underpinning section III of "The Future We Want." The workshops' main contribution was to make clear, from an early stage of the Rio+20 preparatory process, the concern coming from a majority of developing countries that the concept of the green economy could be accepted only if it were defined as a set of tools, policy approaches and national economic models undertaken to achieve sustainable development. The second contribution of the workshops was to make evident not only the game-changing potential behind the idea of a transition to a green economy approach, but also the challenges for its implementation at the national level. In this regard, the workshops were useful in focusing attention on the fact that the transition towards a green economy required structural change in national production structures, and involved a technological revolution, changes in production and consumption patterns, and availability of innovative sources of financing to accelerate innovation and transmission process of technologies in developing countries. The workshops also opened a necessary discussion to clarify the issue of barriers to the implementation of green economy strategies, with potential trade-distorting obstacles being perhaps the most pressing concern raised by developing countries in the dialogue leading up to the Rio+20 Conference. The final concrete contribution of this group of friends was to publicize in New York, thanks to sharing of actual examples, the fact that the green economy was already being implemented in a substantial number of developing and developed countries.

### *Informal discussion group on the Institutional Framework for Sustainable Development*

In early July 2011 a group of approximately 40 ambassadors, led by Macharia Kamau of Kenya, Hasan Kleib of Indonesia and Claude Heller Rouassant of Mexico, convened an informal discussion group on the Institutional Framework for Sustainable Development (IFSD). This group quickly mobilized the attention of ambassadors who felt that with less than 12 months to go before

Rio+20, there was hardly any consensus on what was at stake in the institutional chapter.

This "informal but structured space" allowed them in the course of the second half of 2011 to improve understanding of the reforms that Rio+20 could yield with each of the institutional dimensions of sustainable development and the integration and coherence among and within the three. The overall idea was to work on the basis of the best use of existing structures, through a functional, issue-solving approach (e.g. how can the identified key areas and institutions involved in the IFSD perform better or be strengthened?). Those meetings resulted in a specific analysis and in policy recommendations, some of which eventually made their way into "The Future We Want."

The overarching message of the group was that the UN institutional system was not performing in accordance with its design. Concerned ambassadors felt that the integration of policy-making, implementation and financing of the three pillars of sustainable development was particularly weak. There was also a consensual view that by adopting a more pragmatic or programmatic approach, the Rio+ 20 outcome could significantly improve the way institutions work. It could also increase the benefits of national and local-level implementation by reallocating funds to avoid duplication of activities, clarifying functions, roles and responsibilities of the different parts of the UN system.

On the basis of such analysis, by October 2011, just before the deadline set for national submissions for the preparatory process of Rio+20, a common view started to emerge in New York thanks to the work of this group that member states could adopt a realistic package based on key basic functions of an enhanced IFSD. There was also a good measure of convergence on a few key points that could guide a consensual outcome in this area. The first common understanding was on the need for coherent, coordinated and balanced integration among and within the three pillars of sustainable development. The second common understanding was the importance of refocusing the discussion on IFSD on implementation, and of favouring any measure that would reduce overlaps and increase implementation. As the Rio+20 process evolved in 2012, two more functions were added: enhancing the inclusiveness with regard to stakeholders and improving the science–policy interface.

Right from its origins, the work of this discussion group was supported by an important number of influential ambassadors who proactively hosted workshops and discussions: on the economic pillar, Mourad Benmehidi of Algeria, Jan Grauls of Belgium and Modest Jonathan Mero of Tanzania; on the social pillar, Yanerit Morgan of Mexico and Eduardo Gálvez of Chile; and on the environmental pillar and overall challenges of implementation, Ambassadors Morten Wetland of Norway and Peter Wittig of Germany. This group was also advised by Oscar de Rojas, former director of the UN Financing for Development Office; Ivar A. Baste, former director of the Environment Management Group (EMG); and Patrick Tiefenbacher, then adviser on corporate innovation for the

UNDP. It involved, however, a wider number of discussants and panellists from institutions belonging to the social, economic and environmental pillars.

The function-based approach taken initially by the discussion group gradually progressed towards a more specific tool-based dialogue. While functions and tools remained at the base of all IFSD discussions in New York, specific institutional options did emerge. This discussion group was the only space in New York for ambassadors to discuss, from a cross-regional perspective, options for achieving a balanced integration within the three pillars of sustainable development (vertical integration) such as strengthening existing institutions like ECOSOC or the CSD, or establishing new ones, like a Global Ministerial Forum on Sustainability or a Sustainable Development Council.

The work of this group served to identify the fact that Rio+20 also provided an opportunity to address the environmental pillar, including the three existing options addressed by the UNEP-led Nairobi–Helsinki process (enhancing institutional reforms, enhancing UNEP and establishing a specialized agency such as a World Environment Organization). While there were also discussions looking at alternatives for improving the coherence of the social and economic pillars, the informal character of the discussion group allowed ambassadors to reach the conclusion that the ongoing constraints made reforms in these pillars beyond the reach of the mandate for Rio+20.

### *Non-meetings on the IFSD and the emergence of the high-level political forum*

The seeds of three of the major IFSD outcomes achieved in section IV of "The Future We Want" and through its UNGA implementation via resolutions in 2012 and 2013 (Resolution 66/2013 on UNEP, the GC Resolution 27/1 on the UNGA resolution, the High-Level Political Foundation and the ECOSOC-strengthening resolution) were shaped in informal non-meetings convened between February and May 2012 by the then vice-president of ECOSOC, Ambassador Luis Alfonso de Alba. These key deliverables were the replacement of the Commission on Sustainable Development by the hybrid High-Level Political Forum and the transformation of the Governing Council of UNEP into a United Nations Environment Assembly. They were both introduced in a series of non-papers entitled "An Agenda for the Institutional Framework for Sustainable Development," the last version of which was released in early May 2012.

The basic idea informing the non-meetings on IFSD was that Rio+20 provided a generational opportunity to overcome the artificial separation between the so-called sustainable development agenda, enshrined in Agenda 21, and the development agenda, contained in the Millennium Declaration and the MDGs. The view was that to achieve this outcome, the institutional arrangements had to be geared towards the integration and convergence of the two development agendas.

The key policy implications addressed in those non-meetings, which gathered ambassadors mainly from the bureaux of Rio+20, the CSD and ECOSOC, was that in order for there to be a consolidated development agenda beyond the 2015 timeline for the MDGs, different existing processes needed to converge.

The first contribution of the non-meetings was to build a common understanding on the idea that the proposed Sustainable Development Council, although a low-hanging fruit, had indeed little potential to effectively integrate the social, economic and environmental agendas within the United Nations. And if it were made separate from ECOSOC, the result would only be increased fragmentation and unnecessary overlaps with regard to other bodies and processes, most notably concerning UNEP. If the purpose of Rio+20 was to lay the basis of an integrated global development agenda, logically – and regardless of its conceptual appeal – it could not conclude with the establishment of yet another council.

A second initial conclusion was that the mandate and functions of ECOSOC could be interpreted to cover the environmental pillar, hence making it unnecessary to reform the UN Charter, something that had previously been assumed to be indispensable and argued for over a period of many years. A third initial conclusion was a growing realization that the CSD was unfit to be "strengthened" in order to serve as the space to achieve the integration of the three dimensions of sustainable development and to prepare and periodically review the implementation of an integrated development agenda beyond 2015, joining the MDGs and the SDGs.

It therefore needed to conclude its functions by 2013, a year after Rio+20, in order to allow a transition towards improved institutional arrangements, before another CSD cycle became due. Finally it was agreed that a "door" or "space" needed to be opened between the UNGA and ECOSOC in order to ensure universality and foster policy coherence between the agenda-setting and review functions of both principal organs.

The non-paper by Ambassador de Alba proposed that to achieve the functions and objectives of the IFSD,

> leaders can decide to establish a sustainable development assembly as the high level intergovernmental forum at the United Nations to provide, through the exchange of experiences and focused, dynamic and action-oriented working methods, high-level political guidance, definition and dialogue for the global implementation of the sustainable development agenda; increase system-wide coherence, cooperation and coordination; and address interlinked issues.

The non-paper further proposed that

> to ensure leadership and enhance integration, such a space could be convened jointly at the heads of state and government level between the

*Figure 2.1* Mexican ambassador Luis Alfonso de Alba and Dámaso Luna, Mexican envoy to the Open Working Group on Sustainable Development, during the first round of informal consultations, March 2012 (courtesy of IISD/*Earth Negotiations Bulletin* (www.iisd.ca)).

> General Assembly and ECOSOC every four years in September while ensuring annual ministerial sessions convened by ECOSOC.

It was being mooted that such a

> high level forum should ensure the involvement of civil society and the system-wide participation and support of the UN specialized agencies, funds, programs, International Financing Institutions and other financing entities, the WTO, UN Regional Commissions and the main Conventions and Bodies.

### *Box 2.2* The blueprint of the high-level political forum

The UN requires a common space to achieve the integration of the three dimensions of sustainable development and to prepare and periodically review the implementation of an integrated development agenda. To be effective and durable such a common space will require tools grounded in functions.

### *Political leadership and agenda setting for sustainable development*

Establish a Sustainable Development Assembly/Global Forum on Sustainability as the principal forum for policy definition, policy review, policy dialogue and cooperation for the implementation of an integrated sustainable development agenda;

The Assembly/Forum could be a universal forum convened by the ECOSOC and the General Assembly.

To ensure leadership and enhance integration, such a space could convene at the heads of state and government level, possibly every four years in September, while ensuring annual high-level sessions in which economic, financial, social, environmental and other ministers would follow and advance the implementation of the sustainable development agenda and consider emerging issues.

The Assembly/Forum will be open to all relevant actors engaged in the SD agenda, namely member states, the UN System including the international financial institutions and other financing facilities, agencies, funds and programmes, the three so-called Rio Conventions, and open to the broad participation of civil society and Major Groups.

### *Ensure implementation*

Establish a universal periodic review mechanism to promote and support national and international progress in achieving sustainable development.

Such a periodic and integrated mechanism would consider the main aspects of the implementation of the sustainable development agenda (national, institutional and cooperation).

"An agenda for the Institutional Framework for Sustainable Development," April 2012.

With regard to the future of UNEP, the non-meetings contributed the following proposal, which would emerge as a realistic alternative for all countries after "The Future We Want" was agreed:

> Ensure that UNEP is responsive to all member states by transforming its Governing Council into an Environment Assembly in order to allow equal participation of all member states in its deliberations; to give to such a high

> level intergovernmental body the authority to fulfill the functions and responsibilities defined in UNEP's mandate; to lead and set the global environmental policy agenda and actively involve stakeholders and civil society.
>
> (United Nations, 2012a)

Moreover, it was felt that there was the need

> to increase the number of countries represented in the Bureau in order to reflect its universal character and to facilitate oversight and standing governing functions of UNEP;
>
> To strengthen the linkages with the sustainable development forum/assembly in order to ensure the integration of the three dimensions of sustainable development" and finally to:
>
> agree on more secure stable, additional and predictable financial resources to UNEP from a combination of the strengthening of the UN regular budget and the voluntary funding through the existing Voluntary Indicative Scale.

The non-meetings were not meant to introduce any textual proposals to the process of negotiations but only to foster a better understanding and build consensus in light of the little progress achieved in negotiations.

## International meetings that gave Rio+20 a global impact

The Rio+20 informal preparatory process also involved an important number of international preparatory meetings. This section discusses eight of these meetings, whose purpose was to provide decision-makers with a dedicated opportunity to advance greater commitments on the different areas covered by the conference.

### *Institutional Framework for Sustainable Development, Solo, Indonesia, July 2011*

The government of Indonesia, working with the United Nations Department of Economic and Social Affairs (UN DESA), organized a High-Level Dialogue on the Institutional Framework for Sustainable Development in Solo, Indonesia.

The background paper from UN DESA brought into focus a number of issues that resonated in the meeting and would later play a significant role in the coming year. The first issue was that the institution set up by the first Rio Earth Summit in 1992 to address sustainable development, the United Nations Commission on Sustainable Development (CSD), had certainly had a number of successes but over the years had become inadequate to address the interlinkages of sustainable development and to offer a high enough political forum to move forward the agenda of sustainable development.

### *Box 2.3* Successes of the United Nations Commission on Sustainable Development

1994: The CSD called for the development of "effective legally binding instruments concerning the Prior Informed Consent (PIC) procedure on the importation of chemicals."

1995: The CSD established the United Nations Intergovernmental Panel on Forests.

1996: The CSD outlined the requirements for the establishment of the institutional arrangements for the implementation of the Global Programme of Action for the Protection of the Marine Environment from Land-Based Activities.

1997: A United Nations General Assembly Special Session (UNGASS) called for, by 2002, the formulation and elaboration of national strategies for sustainable development, the establishment of the United Nations Intergovernmental Forum on Forests (IFF) for three years (1997–2000) and the establishment of multi-stakeholder dialogues with governments within the CSD.

1998: The CSD called on UNCTAD, UNEP and UN DESA to help develop a Vulnerability Index for the quantitative and analytical work on the vulnerability of Small Island Developing States and the establishment of a review of voluntary initiatives within industry.

1999: The CSD established an expansion of the UN guidelines on consumer protection to include sustainable consumption. It also established an open-ended informal consultation process on oceans and seas under the UN General Assembly.

2000: The CSD outlined the terms of reference for a new permanent body: the United Nations Forum on Forests.

By the time of the hosting of the informal government "High Level Dialogue on the Institutional Framework for Sustainable Development" in July 2011 in Solo, Indonesia, it was clear, even to the most supportive government or stakeholder group, that the current way of addressing sustainable development in the United Nations had to change. The meeting allowed participants to draw some initial conclusions regarding the experience from the 2003 reform of the mandate and programme of the CSD, which led to an overtly rigid character imposed by the two-year cycle logic, as well as to the loss of all accountability mechanisms within the Commission, since the link between policy decisions and funding was broken. Development ministries were no longer required to report on the funding of the delivery of Agenda 21. The background paper written for UN DESA (Dodds, 2011) offered three options:

- option 1: incremental: addressing politically sensitive issues, having continuity between review and policy years, enhancing the contribution of Major Groups and strengthening the role of the Bureau;
- option 2: overhaul: return to a one-year cycle, restore four to five weeks of annual meeting space for the CSD, ensuring that the mandate of the CSD includes reviewing means of implementation to ensure that development ministries are engaged and creating a high-level segment to deal with new and emerging threats;
- option 3: new institutions: by establishing a Council of the UNGA, as recommended by the Global Sustainability Panel Report, that would conduct a review of previous commitments and assess new and emerging issues.

These three options were also outlined at the same meeting by Martin Khor (2011), executive director of the South Centre, an intergovernmental organization established in 1995. Khor had for much of the previous 20 years been executive director of the Third World Network, which had played a significant role in voicing the views of countries in the South in international sustainable development fora over that period. He raised an issue regarding option 3, which would become central to the discussion over the coming year: should a transformed CSD be an independent council of the UNGA or continue to be a subsidiary body of ECOSOC? The meeting in Solo and the social events around that meeting featured discussions on these options very similar to those under way in the different informal tracks in New York. Nevertheless, the meeting allowed participants to realize that the doubts surrounding the future of the CSD were not only felt within developing countries. European Union countries were facing similar difficulties in reaching a unified position on the future of the CSD. Some of them, particularly Denmark, Finland, Germany and the Netherlands, were openly interested in such reform. Other European countries, such as Liechtenstein, Norway and Switzerland, were also staunch advocates of the transformation of the Commission into a Council subsidiary of the General Assembly.

The deficiencies of the environmental pillar and the proposal for upgrading UNEP into a World Environment Organization (WEO) were also addressed at the Solo dialogue. Its open format made it possible to appreciate the major obstacle to the initiative for transforming UNEP into a specialized agency. Despite the increasing number of countries that expressed their support for the transformation during 2011 and 2012, which at a certain point totalled close to 90 countries – technically, sufficient to adopt a resolution by vote in the General Assembly, provided that most other countries abstained – the proposal had not managed to persuade others, including Canada, China, Russia and the United

States, as well as the countries of the ALBA alliance and several North African and Arab countries, which openly expressed caution regarding the establishment of a new organization. Without these countries, no decision without recourse to a vote would have been possible. However, the lack of agreement on the draft rules of procedure for Rio+20 made it technically impossible to envisage any procedural motion.

These issues were not ignored by the key actors involved, yet their concerted efforts allowed most countries to identify common interests in an enhanced – though not necessarily transformed – UNEP. In fact, in Solo all participating delegations expressed a willingness to substantially review the functions, mandate, resources and status of UNEP within the United Nations. This represented an opening to adopt decisions on international environmental governance that had not been possible at previous conferences such as Johannesburg 2002 or the World Summit of 2005. In Solo, for the first time, the US delegation, which was formed of representatives at a very senior political level, expressed a willingness to entertain reforms within UNEP's Governing Council, such as its enlargement to universal membership. The same was said by other countries concerning UNEP's funding status and linkages to the operational activities of the United Nations, an issue that nevertheless affected the relationship between UNEP and the UNDP.

The outcome from the Solo meeting was the "Solo Message" proposed by the chair of the meeting, who captured the discussion very well when he said:

> At the international level, we need an organization to enhance the integration of sustainable development. Various options were discussed, ranging from an enhanced mandate for ECOSOC and reviewing the role of CSD, to the establishment of a Sustainable Development Council.
>
> (Indonesia, 2011)

Another issue that moved up the agenda during Solo was that of corporate social responsibility. The discussion paper for the meeting reminded governments of the missing chapter 41 of Agenda 21 in 1992, called "Transnational corporations and sustainable development." It had originally been submitted by the United Nations Centre on Transnational Corporations in 1991, and the chapter was very effectively suppressed by lobbying from the US government, the UK government and other governments opposed to any additional regulation on companies in a time when the world's economies were rapidly globalizing. The discussion paper suggested that Rio+20 should agree to develop a convention on corporate social responsibility, which focused on mandatory sustainability reporting for companies listed on the stock exchange.

The final key issue that was raised in Solo was the proposal of Sustainable Development Goals (SDGs). The idea was articulated by Paula Caballero Gomez, the director of economic, social and environmental matters at Colombia's Ministry of Foreign Affairs, and supported by the government of

Guatemala. The original proposal for the SDGs was grounded on the idea that the Millennium Development Goals (MDGs) had played a significant role in focusing the world community. The eight goals were targeted at the developing countries. However, the only universal goal focused on the environment, MDG7, was seen to be very weak. The original proposal for the SDGs advocated reinvigorating the spirit of MDG7 by updating the sectoral chapters of Agenda 21 and of the Johannesburg Plan of Implementation with sectoral up-to-date targets.

Although the original proposal significantly evolved during the following months, the chair's text already reflected the value of this new idea of SDGs: "There was a significant interest on the discussion on the Sustainable Development Goals" (Indonesia, 2011). The chair's text also recognized the likely difficulty of negotiating new goals during the Rio+20 processes when it said, "In the context of goals related, in particular, to the sustainable energy goals advanced by the [Secretary-General's] Advisory Group, there was a feeling that negotiating specific goals would bog down the negotiations" (ibid.). However, the participants of the Solo meeting thought there would be a good chance for an agreement, in principle, to develop a set of suggested areas for Sustainable Development Goals. More importantly, as the United Nations was starting to prepare for the development of new goals through a parallel Secretariat-led process under the follow-up of the 2010 review of the MDGs (see Chapter 1), sustainable development was not then in the framework under discussion. Solo helped put SDGs on the Rio+20 agenda and Rio+20 put SDGs at the centre of the Post-2015 Development Agenda.

### *The Sixty-fourth United Nations DPI NGO Conference: "Sustainable Societies – Responsive Citizens"*

The UN Department of Public Information (DPI) has held an annual UN DPI NGO conference since 1947. In August 2010 the then director of the Outreach Division of the UN DPI, Eric Falt, approached Felix Dodds to ask whether he would consider chairing the 2011 UN DPI NGO conference. Falt – who left the DPI in late 2010 to become Assistant Secretary-General at UNESCO – had revolutionized the outreach work of the UN DPI on a number of fronts. Taking the long-standing NGO conference "on the road" as a peripatetic meeting was one way to reach to a more diverse group of stakeholders.

The sixty-fourth UN DPI NGO conference was held in the former German capital, Bonn, on September 3–5, 2011. The main partners for the conference that worked in conjunction with the DPI included the German government, UN Volunteers (UNV), the City of Bonn, UNEP and UN DESA. The conference represents a genuinely equal partnership between the United Nations and the NGO community. Maria-Luisa Chavez, chief of NGO Relations at the DPI, who co-chaired the preparatory process with Felix Dodds, was in charge of overseeing the conference. The chair of the DPI NGO Committee was

Jeffery Huffines, from the global NGO CIVICUS. Huffines had been chair of the committee since 2008 and therefore brought a wealth of experience and valuable advice to the preparation for the 2011 conference. He went on to play a very significant role in coordinating NGO input to the Rio+20 Conference itself after CIVICUS became one of the NGO Major Groups organizing partners for Rio+20.

The sixty-fourth UN DPI NGO Conference was unique in its approach to influencing Rio+20 because it brought together 2,000 stakeholders from around the world, as well as a number of key governments and all the top UN agencies and programmes. The conference was a mix of high-level panels, workshops, training, capacity-building and networking. The central outcome was to be a chair's text.

In preparation for the conference, a group of experts from around the world came together on a committee, including some of the best leaders from the NGO sustainable development world.[1] They were facilitated by Richard Jordan of the International Council for Caring Communities, a former chair of the UN DPI conference himself, and Sean Southey of PCI Media Impact, a former secretary-general of ICLEI, Local Governments for Sustainability. Just after the Solo meeting, the experts put forward a draft of what would become the basis of the chair's text, in which were a number of the suggestions from the Solo meeting. The UN DPI NGO conference was the first to respond to the idea of sustainable Development Goals and went ahead to propose to governments a list of 17 SDGs. The conference also supported the upgrading of UNEP and the call for a convention on corporate social responsibility, as well as a convention on Principle 10 of the Rio Declaration on Access to Information, Public Participation and Environmental Justice.

That year, 2011, was the first time that the UN DPI NGO conference had ended with a chair's text, as opposed to an agreed declaration; this was a harbinger of how difficult Rio+20 was going to be. The absence of negotiation over the text enabled it to go through the final plenary without any objections. Much of the chair's text was then reflected in NGO submissions to the Zero Draft for Rio+20. Another indication of the changing nature of the relationship between the United Nations, governments and stakeholders was that the briefings produced by the Rio+20 Secretariat included the recommendations from the UN DPI NGO conference.

In a powerful demonstration of the challenge represented by safety and security matters in the modern world, on the final day the venue of the conference, the Maritim Hotel, went under lockdown by UN Security. A hired car going to the underground garage had tested positive for explosive residue. As UN Security had already conducted a comprehensive sweep of the building, they felt it was safer to keep the participants inside in case there was a car bomb outside. The conference had started only a week after a suicide bomber detonated a bomb outside the UN headquarters in Abuja, Nigeria, killing 18 people.

The UN DPI NGO conference, although not attended by a large number of governments, did have a number of UN under-secretaries-general and leading UN staff, and a further 2,000 stakeholder representatives. It could easily have been seen as a "soft target." The excellent work by Bonnie N. Adkins, chief of UN Security and Safety at the International Criminal Tribunal for the former Yugoslavia in The Hague, in overseeing the UN Security team was impressive. She kept the conference organizers on top of developments in a timely manner as they developed. The discovery of the explosive residue traces in the hired car led to the arrest of two men in Berlin several days later who were involved in a German bomb plot (Kulish, 2011).

## *Outreach efforts for the Sustainable Development Goals*

The SDGs emerged as one of the core results of Rio+20 and "The Future We Want," with member states agreeing that a collaborative and negotiated process should guide the preparation of a formal list of goals

> coherent with and integrated into the United Nations development agenda beyond 2015, thus contributing to the achievement of sustainable development and serving as a driver for implementation and mainstreaming of sustainable development in the United Nations system as a whole.
>
> (United Nations, 2012g, paragraph 246)

This approach is to be sharply contrasted with the one followed for the Millennium Development Goals (MDGs), which were developed in 2000 and were largely a product of the international system – the UN organizations working in consultation with the Organisation for Economic Co-operation and Development, the World Bank and the International Monetary Fund – after a series of international conferences, concluding with the Millennium Summit. The MDGs were subsequently agreed by the United Nations' 193 member states and 23 international agencies as one of the sections of the 2000 Millennium Declaration. Once the goals had been adopted, the targets and indicators used to track their implementation were then refined through an inter-agency process culminating in the 2001 report of the UN Secretary-General entitled "Road Map towards the Implementation of the United Nations Millennium Declaration" (United Nations, 2001).

### *Background to the Sustainable Development Goals*

While the ownership of the SDGs proposal ultimately belongs to the relentless political and outreach efforts of its champions, the governments of Colombia, Guatemala, Peru and the United Arab Emirates, there are several antecedents that have to be quoted to make it possible to understand fully the success of the idea. Among these are the process for developing during 2010 the Aichi

Biodiversity Targets, the work during 2010 and 2011 of the Global Sustainability Panel Secretariat and the conclusions of UNEP's *Global Environment Outlook 5* (*GEO-5*).

The question of whether Rio+20 should conclude with the adoption of goals or targets was certainly not entirely new. The initial discourse towards Rio+20 of the Group of 77 or the European Union was very much grounded on the notion that the conference should result in a tangible "sustainable development action framework" or on a "green economy road map" in which all countries should be led to adopt "additional commitments." After all, goal-setting has been the major staple of all UN international negotiation processes.

In the field of the environment alone, associated with MDG7 (on Ensuring Environmental Sustainability), UNEP's Global Environmental Goals (GEGs) initiative has tracked an impressive number of 235 international goals. Eighty-eight of those are contained in legally binding treaties and instruments, while the remaining 147 goals are set in voluntary non-legally binding documents, such as UN resolutions and the outcomes of international UN conferences, like the Millennium Declaration or "The Future We Want." The UN Integrated Implementation Framework also has a similar tool to track hundreds of intergovernmental commitments for the achievement of MDG8 (on the global partnership for development).

While UN member states have been quite successful in setting goals and commitments, they are still far from achieving their implementation satisfactorily. Why? Aside the obvious reasons (among them the lack of financial and technical resources, lack of adequate integrative policies or structural barriers), in many cases implementation has not been helped by the way the international community has set its own development goals and the mechanisms to review and track them. The main lesson learned, whether through the assessment of the eight MDGs or of the 235 GEGs, is that it is not enough simply to set a goal; it has to be the "right goal," with the "right unit of measurement" and the "right institution" to track its implementation and reward its achievement. That was precisely one of the key purposes of *Global Environment Outlook 5* (*GEO-5*), reviewed in Chapter 1. *GEO-5* – the most ambitious and comprehensive multi-stakeholder environmental assessment – concluded with a word of caution regarding the setting of new goals at Rio+20. *GEO-5* warned that countries had failed to fulfil most of the existing goals. Out of 34 major environmental GEGs, by 2012 positive progress could be reported for only 3 (UNEP, 2012).

The message was clear: any kind of goal had to be integrative in nature (balancing social, economic and environmental factors); scientifically informed; endowed with baseline indicators to track its achievements; and provided with an appropriate framework for its review and implementation. This new approach to a "second generation of goals" was tried for the first time in October 2010 in Nagoya, Japan, during the tenth meeting of the Conference of the Parties to the Convention on Biological Diversity, where countries agreed to the "Strategic Plan for Biodiversity 2011–2020," containing 20 headline targets for 2015 or 2020 – also known as the "Aichi Biodiversity Targets." These targets are organized under five strategic goals combining social, economic and environmental aspects. In fact, many of the leading actors behind the negotiations of the Aichi Targets were those behind the development of the final agreement for the SDGs at Rio. For example, the Brazilian ambassador, Luiz Figuereido Machado, led his own country into agreeing to a last-minute compromise for the adoption of the Aichi Targets. Hugo Schally was the European Union lead negotiator at Nagoya and at Rio+20. The same can be said of Paula Caballero, the Colombian proponent of the SDGs; of Dámaso Luna, the Mexican envoy for sustainable development; or of Eidun Edhem, one of the lead negotiators from Norway, whose country also supported many of the outreach efforts for the SDGs. Notwithstanding the appeal that a tangible outcome, from a political perspective, has for the conclusion of an international conference, the consultative, integrative, intergovernmental and scientifically informed experience of the Aichi Targets without any doubt influenced and inspired the SDGs, which at a certain point were also called "Rio Targets."

From a completely different perspective, the Global Sustainability Panel (GSP) Secretariat, led by Janos Pasztor, also worked, independently from the Colombian initiative, on a proposal for Sustainable Development Goals. The

*Figure 2.2* The agreement for Sustainable Development Goals required extensive outreach. Pictured are Farrukh Khan from Pakistan, Hugo Schally of the European Commission, Paula Caballero from Colombia and Franz Perrez from Switzerland (courtesy of IISD/ *Earth Negotiations Bulletin* (www.iisd.ca)).

idea was sufficiently mature by May 2011, when it was discussed at the third meeting of the panel, held in Helsinki on May 16–17, 2011. The summary of that meeting, which was attended by the Presidents of South Africa and Finland and by most panel members, points to the advanced state of discussion on possible goals at that point in time:

> The Panel discussed the possibility of developing long-term sustainable development goals and indicators that could be applicable in the post-MDG period. In view of the time and technical expertise such an undertaking would require, it was agreed that the Panel would consider calling for such goals to be developed, perhaps also indicating some elements that could be included, without getting into details.
>
> (Global Sustainability Panel, 2011)

By June 2011 the first draft version of the report of the GSP already contained some indications of the scope of what would eventually become the core of the process to elaborate the SDGs and to define its relationship with the MDGs. The relevant paragraph of the draft report, which was substantially reduced in its final version, says:

> [T]he world is making progress in reducing poverty, but the progress is uneven and incomplete, and it has to contend with the absolute rise in human numbers. So what should happen? Since the 2010 MDG Summit, discussions on the future of the MDGs and the development agenda after 2015 have gathered momentum. One suggestion is that perhaps the Millennium Development Goals could morph seamlessly after 2015 into something more comprehensive – the sustainable development goals.
>
> (GSP Report, Preliminary Draft of Parts I and II, June 2011)

The GSP worked under the assumption of three alternative models for the development of the SDGs, all of them to be developed in an intergovernmental setting:

1 Keeping the current MDGs while adding up-to date elements to them. For example, in the case of MDG 7 (Ensuring Environmental Sustainability), governments could draw on the experience of the United Nations Commission on Sustainable Development (CSD).
2 Expanding the scope of the current MDGs to fill gaps in the current list and to put the relevance of human rights into more concrete terms. For example, MDG 1 could have the additional target of a global social protection floor.
3 Defining new development models and goals. With their restricted focus on extreme poverty and basic social services, the MDGs reflect the lowest common denominator that could be agreed ten years ago. A more com-

prehensive approach would stress the equal value of the environmental, economic and social aspects of sustainable development, as well as its human rights foundations.

The final version of the GSP report, *Resilient People, Resilient Planet* (United Nations, 2012c), contained an endorsement of a set of "universal sustainable development goals" and a recommendation for their development through an expert mechanism. Its final recommendation read:

> Governments should agree to develop a set of key universal sustainable development goals, covering all three dimensions of sustainable development as well as their interconnections. Such goals should galvanize individual and collective action and complement the Millennium Development Goals, while allowing for a post-2015 framework. An expert mechanism should be established by the Secretary-General to elaborate and refine the goals before their adoption by United Nations Member States.

By the time the GSP report was released, its recommendation reflected – rather than influenced – the current state of negotiations. However, the work of the GSP was essential for "diffusing" among key countries the idea of using Rio+20 to integrate the three dimensions of sustainable development through a limited, universal and transformative set of goals.

### *Major outreach efforts for the SDGs undertaken by Colombia*

In early May 2011, Paula Caballero, Colombia's lead negotiator for the Rio+20 process, with the full support of her minister, María Ángela Holguín, started sharing informally with a small number of key UN delegations in New York her concern that the Rio+20 process was losing direction. The conference was only a year away and the discussions on its main issues – the future of the CSD, a green economy road map, a thematic action framework, a technology facilitation mechanism and a renewed mandate for UNEP – were all at a standstill. The idea was to think out of the box and propose something new that would restart the engine of the Rio+20 preparatory process. That was the original aim behind the Colombian proposal of the SDGs.

The main outreach efforts were undertaken by Colombia, Guatemala, Peru and the United Arab Emirates, whose collective influence in the development and socialization of the proposal were as follows:

- May 2–13, 2011: nineteenth discussion in the sidelined session of the Commission on Sustainable Development (New York);
- July 19–21, 2011: High-Level Dialogue for the Institutional Framework for Sustainable Development (Solo, Indonesia);
- September 7–9, 2011: Economic Commission for Latin America and the Caribbean (ECLAC) Regional Consultation (Santiago, Chile);

- November 4–5, 2011: informal consultations on the SDGs proposal organized by the government of Colombia and ECLAC (Bogotá, Colombia);
- January 22–24, 2012: retreat on "SDGs, Rio+20 and the Post-2015 Development Agenda" (Tarrytown, New York);
- February 2012: proposal formally introduced by Colombia and co-sponsors for discussion within the Group of 77 (New York).

The initial Colombian Concept Note, advanced for the first time in an open format in Solo (Colombia, 2011), was repeatedly revised as the discussions evolved and stakeholder opinion was canvassed and considered. It anticipated that "these Sustainable Development Goals would translate the Green Economy/Sustainable development debate into tangible goals which would focus the broad debate at a practical level and enable the preparatory process to productively address key issues." The goals would number no more than ten and would be guided by the principles contained in Agenda 21, common but differentiated responsibilities and the Johannesburg Programme of Implementation. It was then projected that the goals, which were intended to "combat poverty," would "complement the existing MDGS, be easily communicated, voluntary, universal, time bound, organized by thematic areas, contain quantitative targets, and a suite of aspirational indicators; and such goals should galvanize individual and collective action" (UNECLAC, 2011). As proponent, Colombia envisaged an indicative list of goals as follow:

- food security, production, access and nutrition;
- integrated water management for sustainable growth;
- energy for sustainable development;
- sustainable and resilient cities;
- healthy and productive oceans;
- enhancing the capacity of natural systems to support human welfare;
- improving efficiency in resource use;
- enhancing employment and livelihood security.

Colombia – as was also the trend among Global Sustainability Panel members – held the view that the SDGs "must be defined through targeted consultations and deliberations by member states and not negotiated" (Colombia, Peru and the United Arab Emirates, 2012). This approach was ultimately rejected by member states, which settled for the compromise of establishing an open working group comprising 30 representatives as the apex of "an inclusive and transparent intergovernmental process on sustainable development goals that is open to all stakeholders, with a view to developing global sustainable development goals to be agreed by the General Assembly" (United Nations, 2012c). However, in time, much of the core elements suggested by Colombia would be embraced by all stakeholders.

**Box 2.4 Paula Caballero**

Paula Caballero, director of the Department of Economic, Social and Environmental Affairs of the Ministry of External Affairs of Colombia, led the charge on the Sustainable Development Goals for her country as proposer of the idea that a set of SDGs should be one of the tangible outcomes of Rio+20. Caballero worked tirelessly to achieve her country's objective between the formal launch of the concept in July 2011 and the agreement on an Outcome Document at Rio on May 18, 2012. In those 11 months, Caballero convened numerous meetings, often with the active involvement of her deputy minister, Patti Londoño, and participated in meetings hosted by others. She conducted several one-on-one consultations and engaged with a myriad of stakeholder groups to promote the SDG concept and the contents of a possible list, and thus to obtain agreement on a process. When division among stakeholders threatened failure to adopt the SDGs as part of the outcome, she redoubled her efforts. Her commitment and attitude, her high level of engagement and savviness, and the willingness to make changes to the original concept as member states' and stakeholders' perspectives became known, showed that Colombia had no bias and was not trying to "sell" something to member states. These were all instrumental to achieving a success with which Caballero must be credited.

After its initial proposal, Colombia took its efforts to the next regional preparatory meeting for Rio+20, the ECLAC-sponsored preparatory meeting for the Latin American and Caribbean region held on September 7–9, 2011, at ECLAC's campus in Santiago, Chile. In furtherance of Colombia's thrust to build momentum on the SDGs, the following announcement was made at the meeting:

> By a note dated 5 August 2011 addressed to the Executive Secretary of the Economic Commission for Latin America and the Caribbean, the Ministry of Foreign Affairs of Colombia presented the proposal entitled "Rio+20: Sustainable Development Goals" for consideration in the framework of the Latin American and Caribbean Regional Meeting Preparatory to the United Nations Conference on Sustainable Development.
>
> The secretariat is accordingly transmitting herewith the text of this proposal to the participating countries for consideration.
>
> (UNECLAC, 2011)

The Chile meeting did not reach consensus on the SDGs or on any of the themes or objectives of Rio+20, which were actually the agenda items of the meeting. At its conclusion the meeting agreed on the contents of a Chair's Summary, which included conclusions, a report of the meeting and interventions. In the failure to reach agreement on the Rio+20 objectives and themes, the outcome of this meeting was consistent with those of the other regional meetings. However, the apparent absence of support on the SDGs was the first real signal that the concept might not have a smooth passage as the preparatory process continued. The delegates felt that it was too early to say anything definitive on the SDGs concept and possible list. They did not want to bind themselves without consultation with their capitals. Also, they considered the concept to be too new, the relationship to the MDGs too unclear, the method for arriving at the list of goals was perceived to be surrounded by too much uncertainty and the process for arriving at that list to be too ill-defined. In the weeks that followed, the idea of the SDGs would build momentum and then take root but it was not initially received with universal acceptance or endorsement.

Following the ECLAC meeting, Colombia was present and ensured that similar "introductions" and debates took place at every formal or informal meeting of the Preparatory Committee, regional meeting and virtually all multi-stakeholder dialogues which presented the opportunity to further debate the SDGs idea. It was the willingness to engage, to listen and amend their concept (Colombia, Guatemala and Peru, 2011) on the basis of stakeholders' concerns that helped Colombia break through the initial resistance and garner enough acceptance for the SDGs to become one of the major achievements of "The Future We Want." Initially, however, the ride was rough, despite the eventual almost universal convergence around the concept of the SDGs.

Some of the concerns about the SDGs raised at the ECLAC meeting continued to dominate the discussion as the Rio+20 preparatory process progressed, almost right up until the conference itself. A number of points of controversy and divergence framed the divide in the dialogue. Not all countries supported the idea of SDGs, as some were still uncertain about the position of the SDGs in relation to the MDGs, which they did not want to see abandoned. Another concern was whether the list should be negotiated and concluded before Rio+20, or whether some sort of consensus list could be reached without extensive or acrimonious negotiations. Moreover, there was indecision over what process should be used for concluding the list. Division also existed over whether only the concept of SDGs should be agreed at Rio+20, leaving the actual list and the process for reaching it to be settled after the conference. There was concern by some that the SDGs idea would flounder if there was too much delay and it was left until after Rio+20. These issues became such a sore point that there was a time at which it seemed as if the concept would not make it into "The Future We Want."

The turning point came on November 4–5, 2011, when Colombia hosted a workshop in Bogotá dedicated to discussing the SDGs and taking the concept and process forward. The meeting was held under Chatham House Rule and no formal minutes or report were issued[2] because, according to the host, the intention was to provoke "an open and creative exchange of views among a representative group of member states, as well as UN agencies and NGOs." Pursuant to that objective, countries from every region – developed and developing, industrialized and agrarian, large and small – as well as NGOs from these countries and UN agencies were all invited and participated in the meeting, including two of the authors of this book. Colombia was a generous and gracious host, and with Vice-Minister Patti Londoño and Paula Caballero in the chair, gently guiding discussions without attempting to impose Colombia's own views, the country got its wish of an open, rich and varied discussion on the principles and content of the list of SDGs and an initial definition of a possible process for getting the Goals into "The Future We Want."

Several points of discussion would shape the subsequent debate. There was a general view that the "MDGs should not be undermined" but should "continue their planned trajectory to 2015," although there should be "broad coherence between the existing MDGs and the proposed SDGs," which "could be useful in helping to define a post-2015 agenda." It was felt that the focus of the SDGs should be "major global challenges." Yet, in recognition that there was no "one size fits all" formula that could be applied, some degree of flexibility had to be built into the goals. This attempt at reconciling the relationship between the MDGs and SDGs was important because this issue constituted one of the major hurdles perceived by member states and other stakeholders in coming to agreement on the concept of SDGs.

The workshop decided that "the SDGs are meant to focus on key global challenges that are critical for ensuring the well-being of people everywhere, including through sound stewardship of the natural resource base and services on which so many economies depend." Towards that end, participants at the workshop were of the view that a number of thematic areas were appropriate for inclusion on a list of SDGs:

- food security;
- energy access, including access to renewable resources;
- water, integrated management and sanitary issues;
- oceans and fisheries;
- sustainable human settlements (cities);
- biodiversity and forestry;
- employment;
- women and youth.

As envisaged by participants, a number of guiding principles would underpin the SDGs. They would be

> universal, relevant for all countries, based on Agenda 21 and the Johannesburg Plan of Implementation, forward looking and build upon existing agreements and principles, be fully participatory, be an innovative and bold response to

all three dimensions of sustainable development, should be launched at Rio+20, where a mechanism for developing the list using expert advice and guidance should be articulated.

(Colombia, Peru and the United Arab Emirates, 2011)

With this fillip, the Colombian team, combined with Guatemala and the new additions of Peru and the United Arab Emirates, persevered in their promotion and discussion efforts right through the negotiations. Their efforts were rewarded with success at Rio+20, where member states agreed that SDGs were part of a global sustainable development agenda and made them feature strongly in "The Future We Want." It is a tribute to the work of Colombia that "The Future We Want" contained both principles and language found in its initial proposal and later revisions.

### *The Bonn 2011 Conference: the water, energy and food security nexus*

As mentioned previously, the German government showed considerable leadership within the European Union by investing in additional meetings on topics that needed it. Germany has a rich history of putting together some of the best and most impactful conferences over the past decade. The two most important ones were the Bonn Water Conference in December 2001 and the Bonn Energy Conference in 2004. The Bonn Water Conference is still today commonly viewed as the best example of stakeholder dialogue with government ministers. It was the first time ministers debated with stakeholders on the issue of corruption in the water industry. Also during the dialogue, one of the UK water companies, Severn Trent, announced the introduction of a whistle-blower scheme within the company to address the issue of corruption. In the coming years, this example was followed by most of the other water companies that attended the Conference. The Bonn Conference of 2001 also produced the so-called Bonn Keys (Germany, 2001). These five keys were used to address the water challenges of the future and were a substantive input to the World Summit on Sustainable Development (WSSD) in 2002.

The 2004 Energy Conference, also a success for stakeholder engagement, set up REN21[3] to facilitate knowledge exchange, policy development and joint action towards a rapid global transition to renewable energy. This ultimately led to the creation of the International Renewable Energies Agency (IRENA).

Building on this strong tradition, the German team built on those that had put together the previous two successful conferences: Dr Manfred Konukiewitz, deputy director-general of the German Federal Ministry of Economic Cooperation (BMZ), with support from Franz-Josef Batz, team leader for International Water Policy and Infrastructure (GIZ), and Dr Fritz Holzwarth, the German Federal Ministry of the Environment (BMU)'s deputy director-general for water management, while adding Imke Thirm as the head of the 2011 Secretariat for the Conference.

The German government utilized the Stockholm World Water Week in 2009 and 2010 to map out what would be the most useful contribution to Rio+20. It chose to address something that had been near-impossible during the CSD sessions: the interlinkages between the key issues of water, energy and food, all of which are now clearly adversely impacted by the consequences of climate change.

The approach to this "nexus" was very similar to what has been done for the two previous conferences in Bonn. As before, the organizers brought together a significant advisory group, chaired by Uchi Eid, vice-chair of the UN Secretary-General's Advisory Board on Water and Sanitation (UNSGAB) and a former Parliamentary Secretary for the German Federal Ministry of Economic Cooperation (BMZ), and Albert Butara, the former Minister of Energy and Water of Rwanda. The conference planners subcontracted the background paper to the Stockholm Environment Institute (SEI) and they assigned Stakeholder Forum the work of designing how to engage stakeholders in the conference, as they had done for the two previous conferences.

The SEI produced a compelling Background Report. In the area of water, there were an estimated 900 million people without access to safe drinking water and over 2.5 billion people without access to sanitation. In the area of energy, 2.5 billion people were without access to modern forms of energy and 1.5 billion people without access to electricity. Additionally, 1 billion people were found to be suffering from hunger. The report recognized that there were a number of challenges and tightening constraints such as pollution growth, increased urbanization, growing consumption by an increased middle class and of course the impacts of climate change. These factors, the report's authors estimated, will have a number of impacts:

- Agricultural production will have to increase by about 30–50 per cent by 2030.
- Energy demand will increase by 40 per cent by 2030.

For water, this will mean that by 2030 there will be a 40 per cent gap in water resources between availability and demand. This was the background to the first meetings of the Steering Committee as its members focused their contributions toward an agenda for the Bonn 2011 Conference, "Water, Energy and Food Security Nexus – Solutions for the Green Economy." The conference, scheduled for November 16–18, 2011, was coming after the date set for submission of contributions to the Zero Draft of Rio+20. However, as was done for the Monaco Conference on Oceans, which would take place even later than the Bonn Conference, UN DESA kept a space open for these very important submissions to input their contributions to the Zero Draft.

Just prior to the Bonn Nexus Conference, the World Economic Forum brought out a report entitled *Water Security: The Water–Food–Energy–Climate Nexus*. In its introduction, Indra Nooyi, chair and CEO of PepsiCo, said:

> Water sits at the nexus of so many global issues . . . including health, hunger and economic growth. And sadly, water scarcity takes its greatest toll on society's least fortunate. I am absolutely convinced that the only way to measurably and sustainably improve this dire situation is through broad-scale collaborative efforts between government, industry, academia and other stakeholders around the world.
>
> (quoted in World Economic Forum, 2011)

The outcome from the three-day conference was a co-chairs' summary, which identified a number of key issues:

- Access to water, sanitation, energy, and food is compounded by growing challenges to resource availability, management and sustainability.
- The water, energy and food sectors are connected in important ways, and actions in one sector have the potential to either help or harm the other two. Disconnected approaches and silo thinking are more likely to make matters worse.
- Based on a better understanding of the interdependencies across water, energy and food challenges, the nexus approach provides a more comprehensive base for allocating scarce resources (Bonn, 2011).

The Bonn Nexus Conference recognized how important it was to move this agenda forward and the need to identify the forces that drive the adoption of nexus thinking. It helped to identify mutually beneficial responses and developed an informed and transparent framework for determining and resolving trade-offs between demand and sustainability. Rio in 1992 recognized that in order to carry out the systematic changes that were necessary, institutions would need to change and people would have to build their capacity to address the new challenges. By Bonn 2011 we were returning to a similar narrative, and if the changes were to be successful, greater innovation, cooperation, experimentation and significantly more implementation would be required.

### *The Monaco Workshop on Sustainable Use of Oceans in the context of the green economy and poverty eradication*

In mid-2011 the Principality of Monaco, which has had a long-term interest in oceans and ocean policy, decided to organize a workshop on oceans as part of its contribution to Rio+20. It engaged Stakeholder Forum for a Sustainable Future to help coordinate the meeting and work on the final policy recommendations.

The workshop was led very effectively by Monaco's ambassador to the United Nations, Isabelle F. Picco, and by Valérie Bruell-Melchior, First Councillor. They were supported by Jan-Gustav Strandenaes and Kateryna Wowk of Stakeholder Forum. It focused on three areas: food security and safety, energy and tourism, and their related governance. The Monaco Message, along with the efforts of the group of friends of oceans, played a significant role providing content for the section on oceans and seas of "The Future We Want."

This was to be the only informal meeting in the Rio+20 process that all the major leaders of the process were to attend together. The main participants were the UN secretary-general of Rio+20, Sha Zukang; the two executive coordinators, Elizabeth Thompson and Brice Lalonde; the co-chairs of the Rio+20 Bureau, Ambassadors John W. Ashe and Kim Sook; and Luiz Alberto Figueiredo Machado, Under-Secretary for Environment, Energy, Science and Technology of the Ministry of External Relations of Brazil. The Monaco Message, in addition to endorsing the idea of Sustainable Development Goals, highlighted the call for developing an SDG on the "sustainable development of the oceans, and the importance of the blue economy." Such a goal would

> address all the sectors that rely on ocean resources and space but urgently need to be integrated and made sustainable to continue the provision of their economic, social and environmental services; and to raise the living standards of deprived communities.
>
> (Monaco, 2011)

The second contribution from the Monaco workshop was the call "in the era of global governance" for Rio+20 to "initiate an international agreement on areas beyond national jurisdiction" and "to establish, in order to achieve synergy among all actors and to enhance governance across sectors, a mechanism to achieve greater effectiveness and coherence in the United Nations system" (Monaco, 2011).

### *Annual spring meeting of the World Bank Group with finance ministers, April 2012*

The World Bank Group focused its Rio+20 engagement around four main areas: inclusive green growth; natural capital accounting; oceans, and sustainable energy for all. The World Bank Group supported the discussion around the green economy by offering expertise to assist developing countries to grow their economies in sustainable and innovative ways, as well as by providing innovative financing instruments and partnerships through which the public and private sectors could fund activities, along with advice on metrics to measure progress. Their contribution to the Rio+20 agenda on the green economy was guided by their capacities in a number of sustainable development sector strategies, including agricultural and rural development, energy, infrastructure, information and communication technologies, water, oil and gas, transport, urban, social development, manufacturing, mining, and environment.

On natural capital accounting, the Bank's main objective was to use the renewed momentum given to the System of Environmental-Economic

Accounting (SEEA) by its adoption in February 2012 by the UN Statistical Commission. The SEEA is the international standard for accounting the contribution of natural capital to national income. Its development represents a critical tool for any green economy strategy, as well as for developing adequate metrics that will allow accounting for national wealth beyond gross domestic product (GDP).

On oceans, the World Bank articulated its efforts around a Global Partnership on Oceans (GPO) that was launched in April 2012 in conjunction with the world's pre-eminent ocean-focused organizations, combining knowledge with finance for viable, innovative solutions to connect stakeholders, support knowledge-sharing and meet national demands for improved marine resource management – in particular, on rights-based fisheries management, increasing marine protected areas, improving coastal zone management and reducing marine pollution. This particular work on oceans was a priority for the outgoing World Bank president, Robert Bruce Zoellick.

On April 13, 2012, at the annual spring meeting of the World Bank Group with finance ministers, the Bank held a breakfast meeting specifically to discuss Rio+20: its themes, objectives and the global agenda on sustainable development. It was the first time that the president of the World Bank and the Secretary-General of the United Nations had together presided over a meeting, and the first occasion on which the Bank's meeting with finance ministers would have a session dedicated to issues of sustainable development. Its purpose was described as being to "discuss inclusive green growth and stewardship of the Earth's natural resources, and how to achieve concrete outcomes at the UN Conference on Sustainable Development" (IISD Reporting Services, 2012). The World Bank communiqué issued after the meeting highlighted the "important signal" sent by the meeting "about the need for global partnership to advance the demanding agenda" of "inclusive, green growth in the context of poverty reduction and sustainable development, natural capital accounting and oceans, feeding into the Rio+20 and G20 processes" (International Monetary Fund, 2012). The meeting would also set the stage for the World Bank to launch, as part of its Rio+20 initiatives, its Natural Capital Accounting Project within 50 countries. After the April meeting with finance ministers, the Bank also hosted two further meetings on important Rio+20 themes: a meeting on its new Global Partnership on Oceans and a second meeting on sanitation and water.

### *Planet Under Pressure*

The global science community played a very significant role in the Rio+20 processes, as it had done in the original 1992 conference. Its input to Rio+20 culminated in the Planet Under Pressure conference in March 2012. The conference was organized by the major science platforms the International Geosphere-Biosphere Programme (IGBP), DIVERITAS, the International Human Dimensions Programme on Global Environmental Change (IHDP), the World Climate Research Programme and the Earth System Science Programme. The conference was attended by over 3,000 of the world's leading experts, who gathered to discuss the global challenges and offer new, science-based solutions.

In 2009 a group of environmental scientists led by Johan Rockström of the Stockholm Resilience Centre and Will Steffen of the Australian National University proposed a framework of "planetary boundaries" or tipping points. Through it, they attempted to define what they called a "safe operating space for humanity." Their work defined nine boundaries (Figure 2.3), and they found that three of these boundaries had been crossed: climate change, biodiversity loss and the nitrogen cycle boundary. A debate emerged over which expression was to be preferred, "tipping points," "planetary boundaries" or "critical thresholds." Many rapidly industrializing developing countries, as well as other industrialized ones, have followed a cautious approach towards these notions, particularly the idea of recognizing "boundaries" to their developmental efforts. Despite this opposition, Brazil, China, India, Indonesia and Mexico have used and agreed to the idea of "critical thresholds," specifically in the context of the "Summary for

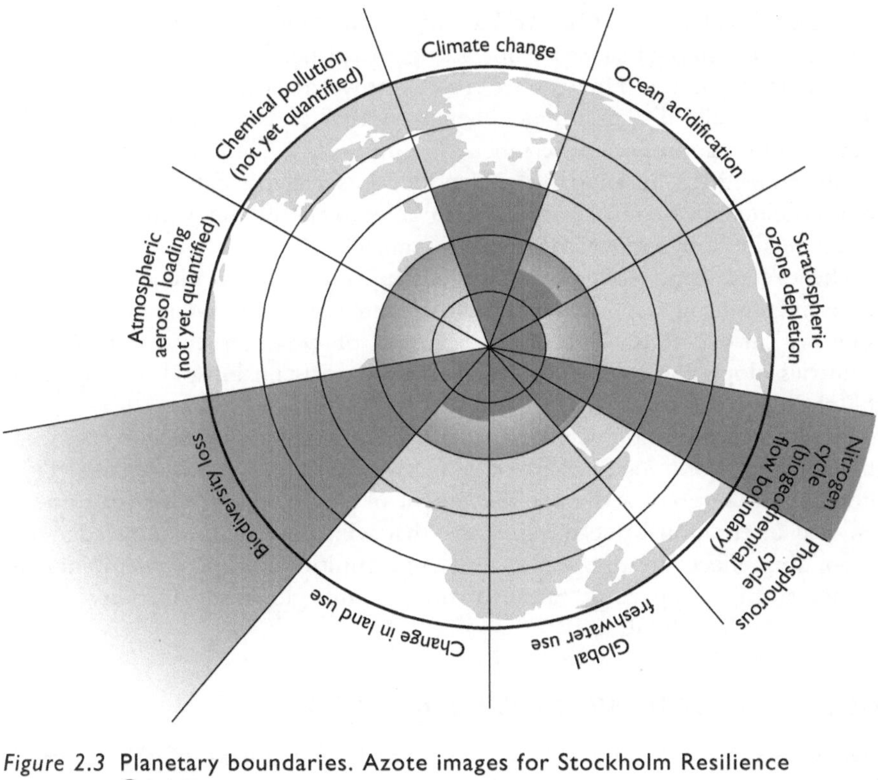


*Figure 2.3* Planetary boundaries. Azote images for Stockholm Resilience Centre

Policy-Makers" of the *GEO-5* report (as explained in Chapter 1). However, none of the three expressions appears in "The Future We Want."

The Stockholm Resilience Centre and the Australian National University published their findings in full in September 2009 in the journal *Nature* (Rockström *et al.*, 2009) which captured considerable press interest and in turn reopened the debate on the boundaries. It was probably not surprising that when the Rio+20 processes were starting, this would become a significant backdrop to the negotiations. The planetary boundaries approach indicated that the developed countries were "leaving a mess," and in order not to breach the boundaries, developing countries would need to modify their development plans so they did not make the same mistakes as their predecessors. This was not an easy message to convey to billions of people who were trying to get out of poverty. In fact, it fed one of the deep concerns of the Group of 77 and China: that the "green economy" concept could be used to rein in the development policies and prospects of the developing world.

Hence, by itself the planetary boundaries approach was unlikely to elicit a positive reaction from many developing countries. Recognizing this, Kate Raworth, a senior researcher at Oxfam, came forward with what would become known as the "Oxfam doughnut" (see Figure 2.4) in a paper called "A Safe and Just Space for Humanity: Can We Live within the Doughnut?" (Oxfam, 2012). The paper recognized that we are now living on a planet that has limited resources and asked the question "How can we ensure that we eradicate poverty and achieve prosperity for all?" To build up a coherent social foundation, Raworth crowd-sourced information from governments' submissions to the Rio+20 Zero Draft in November 2011. In doing so, she identified 11 social foundations that also had to be considered when making policies in order to address the twin objectives of poverty eradication and environmental sustainability.

The joint challenges of ensuring that all humanity progresses in a safe and just space and ensuring that planetary boundaries are respected were identified as an issue within the Rio+20 process regardless of whatever economic model countries adopt. The High-Level Panel on Global Sustainability also recognized this challenge (Oxfam, 2012).

Planet Under Pressure flagged that the global science platforms recognized the need to work together to offer clearer advice to policy-makers. This provided the setting for the announcement of "Future Earth" as a major new ten-year international research initiative that will "develop the knowledge for responding effectively to the risks and opportunities of global environmental change and for supporting transformation towards global sustainability in the coming decades" (ICSU, 2012).

## *SIDS' launch of SE4All, Barbados, May 2012*

Five weeks prior to the start of the high-level segment of the Rio+20 Conference, on May 8–9, 2012 the Sustainable Energy for All (SE4All) initiative

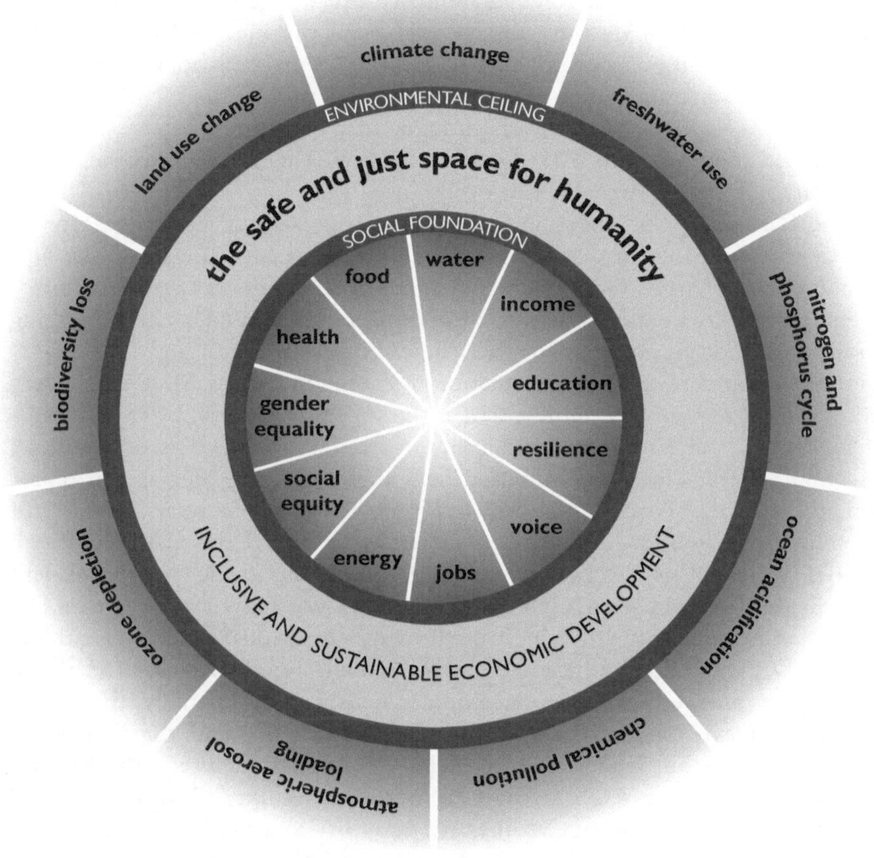


*Figure 2.4* The Oxfam doughnut (source: Oxfam, 2012).

was launched for the Small Island Developing States (SIDS) in Barbados. No one could have known at the time that this would become an important milestone in the preparatory process for Rio+20. There had been SE4All launches in various geographic locations to generate interest and support among politicians, their technocrats, the business sector and other stakeholders. One of the two executive coordinators of Rio+20, Elizabeth Thompson (one of the authors of this book), was a SIDS national who held the view that, given the peculiar vulnerabilities and challenges in SIDS and their inordinately high energy costs, which were among the highest in the world, in addition to issues on access and efficiency and their potential to be models in renewable energy use there should be a specific launch of the SE4All initiative for SIDS. Thompson also advanced the view that it should be held in Barbados, given that it was the place of birth of the 1994 Barbados Programme of Action for SIDS and that the Barbadian prime minister, Freundel Stuart, was serving on the Global Sustainability Panel. Kandeh Yumkella, director-general of UNIDO and UN co-chair of SE4All, immediately embraced the idea and the Barbados government readily agreed to host it. UNIDO then collaborated with the UNDP and its country office in Barbados in order to have the UNDP resident coordinator, Michelle Gyles-McDonough, arrange the organization of this meeting. In addition to the SIDS launch of SE4All, Rio+20 was the other major item on the meeting's agenda.

The Barbados SIDS meeting sharply contrasts with all previous Rio+20 preparatory meetings in that it was opened by the prime minister of the host country, who pledged support for SE4All, stressing his government's commitment to pursuing a national sustainable energy policy and strategy, and affirming the intention of his government to include the private sector and civil society in the development and implementation of energy initiatives. SIDS from every geographical region were represented at the very highest level by prime ministers, deputy prime ministers and ministers from a cross section of ministries including energy, the environment, development, planning, permanent representatives of the United Nations and very senior technocrats. Credit for mobilizing such high-level participation has to go to the Barbados-based UNDP team led by Michelle Gyles-McDonnough and the deputy resident coordinator, Stein Hansen. Gyles-McDonnough was herself a SIDS national. Also in attendance at the meeting were several international agencies, including the UNDP, UNIDO, the International Renewable Energy Agency (IRENA), UN Water and UN DESA; development institutions, namely the Inter-American Development Bank (IDB), the World Bank and the Caribbean Development Bank; agencies dealing with SIDS issues, including the Alliance of Small Island States (AOSIS), the Caribbean Community (CARICOM) and the Organization of Eastern Caribbean States (OECS); a number of other SIDS agencies; representatives of donor countries and their agencies such as the German Society for International Cooperation (GIZ); the European Union; consultants; private-sector representatives; NGOs; diplomatic representatives; and, significantly, the Rio+20 host country, Brazil, sent an ambassador.

Given this level of representation, the meeting had the authority to make decisions and consequently recorded several firsts in the Rio+20 preparatory process. The meeting:

- went beyond issuing a Chairman's Statement and actually issued a Declaration: the Barbados Declaration on Achieving SE4All in SIDS (SIDS Energy for All, 2012);
- announced clear commitments by a group of UN member states to the UN Secretary-General's SE4All initiative and the Rio+20 themes and objectives;
- supported the establishment of the green economy as beneficial to SIDS in achieving sustainable development;

- announced support for a specific mechanism for achieving energy goals in the form of SIDSDOCK, which has been described as "an initiative among member countries of the Alliance of Small Island States (AOSIS) to provide the Small Island Developing States (SIDS) with a collective institutional mechanism to assist them transform their national energy sectors into a catalyst for sustainable economic development and help generate financial resources to address adaptation to climate change" (http://sidsdock.org/);
- identified a list of energy commitments and strategies to be pursued by SIDS;
- supported renewable energy as critical to an energy policy and strategy and as part of the macro-development policy framework of SIDS.

At the outset, the Barbados SIDS SE4All launch was stated to have four broad objectives: establishing awareness of energy conditions in SIDS; soliciting SIDS perspectives on SE4All; articulating the support of SIDS for SE4All being located within the three pillars of sustainable development; and building awareness in governments and governmental institutions for policy formulation and implementation on energy market regulation to attract investment in sustainable energy, fostering recognition of the importance of identifying areas in which energy efficiency can be doubled and renewable energy sources developed.

In addition to SE4All, its targets and commitments, the meeting discussed issues in relation to the SDGs and their global monitoring; the objectives and themes of Rio+20; the energy outlook for SIDS; the creation of an enabling environment for sustainable energy policies and transitions, achieving the three goals of SE4All; strategies to take SE4All, SIDSDOCK and financing energy initiatives forward, and the review of feasibility studies in renewable energy; and a Knowledge Fair. Last but not least, it was agreed to lobby for the hosting of a SIDS Rio+20 follow-up conference. This would be the successor to the UN Conference on Small Island Developing States, which took place in Barbados in 1994 and produced the development road map for SIDS, known as the Barbados Programme of Action (BPOA). The SIDS would achieve this goal, and agreement to such a conference was specifically included in "The Future We Want."

## The Barbados Declaration on energy issues

The Barbados Declaration was a comprehensive document covering a number of energy issues. The declaration:

- reaffirmed the Barbados Programme of Action (BPOA) and the Mauritius Strategy of Implementation (MSI);
- urged development partners to provide, in a timely and predictable manner, financial and technical support to ensure successful implementation of the BPOA and the MSI;

- urged the achievement of SE4All in SIDS, including via the provision to all households with access to modern and affordable renewable energy services while eradicating poverty, safeguarding the environment and providing new opportunities for sustainable development and economic growth.

It also identified a number of challenges:

- the exacerbation by the global crisis in food and finance of already severe peculiarities of SIDS, including increased natural disasters, and environmental challenges;
- the slow pace of international action to address rising carbon emissions and climate change impacts;
- the extreme economic vulnerability of SIDS and their high exposure to volatile oil prices caused by their dependency on imported oil;
- the increased cost of fossil fuels, which has resulted in the diversion of financial resources from climate change and other development activities, as "a major impediment to achieving sustainable development and eradicating poverty."

The declaration identified a number of opportunities:

- learning lessons from the BPOA;
- enhancing cooperation among SIDS;
- capitalizing on the geographical location of SIDS, which makes them suitable for commercially viable options for renewable energy: wind, solar, geothermal, and ocean energy;
- using their great potential for sustainable and renewable energy for energy security, for building resilient economies as well as for energy export to other SIDS.

The declaration noted the requirement of SIDS for appropriate technical and financial resources in capacity-building and technology transfer. It also highlighted the strong interdependence between sectors such as energy, food, water and agriculture and the potential for eradicating poverty, improving health and ensuring meeting human needs. It also recorded a number of commitments:

- to transition to a modern, affordable, efficient renewable energy sector;
- to call on the UN system and development institutions to provide "adequate predictable financial support and investment";
- to build capacity and ensure technology transfer;
- to develop national, regional and inter-regional energy policies, plans and strategies to address the special vulnerabilities of SIDS.

The outcome of this meeting showed tremendous leadership and resolve by the SIDS within the context of the Rio+20 Conference process. The Barbados

Declaration represented an exceptional document in the preparation for Rio+20, to the extent that it contained an annexe with a detailed list of commitments made by SIDS to energy and the SE4All initiative.

## Stakeholders' preparations and their growing influence in the policy definition process

It is perhaps difficult to appreciate how far the involvement of stakeholders in the global processes has come over the past 20 years. It was the Brundtland Report in 1987 that set the train in motion. In the preparation for the World Commission on Environment and Development (otherwise known as the Brundtland Commission), a set of hearings around the world was organized between Commission members and stakeholders. On the Commission and its Secretariat there were a number of key people who would play a significant role in advocating and lobbying for the involvement of stakeholders in the 1992 Rio Conference and in the coming years.

It was at the Earth Summit in Rio in 1992 that stakeholders started being allowed to attend government negotiations and be on government delegations, with some 150 stakeholders recorded as members of governmental delegations. Agenda 21 was the first UN document to recognize rights and responsibilities of stakeholders, and it identified nine groups of interest, which it called "Major Groups." Agenda 21 said:

> One of the fundamental prerequisites for the achievement of sustainable development is broad public participation in decision-making. Furthermore, in the more specific context of environment and development, the need for new forms of participation has emerged. This includes the need of individuals, groups and organizations to participate in environmental impact assessment procedures and to know about and participate in decisions, particularly those which potentially affect the communities in which they live and work. Individuals, groups and organizations should have access to information relevant to environment and development held by national authorities, including information on products and activities that have or are likely to have a significant impact on the environment, and information on environmental protection measures.
>
> Any policies, definitions or rules affecting access to and participation by non-governmental organizations in the work of United Nations institutions or agencies associated with the implementation of Agenda 21 must apply equally to all major groups.
>
> (United Nations, 1992)

Rio identified its Major Groups to be: (1) business and industry; (2) children and youth; (3) farmers; (4) indigenous peoples; (5) local authorities; (6) NGOs;

(7) the science and technology community; (8) women; and (9) workers and trade unions. As Chip Lindner, executive director of the Centre for Our Common Future, said:

> international institutions and national governments must become increasingly more accountable and responsive to the views and expectations of the world's people as a whole. Indeed, it means that as we approach the next century we must move even further in the direction of global democracy.
> (Lindner, 1993)

Over the next 20 years, this formula was used by regional and local councils developing their local Agenda 21s, as much as by national councils and commissions on sustainable development elaborating their national strategies on sustainable development, and increasingly also by UN agencies and programmes for their engagement with stakeholders.

Since then, there have been many experiments to see how best to integrate stakeholder views into the global process from stakeholder dialogues, to hearings, to panels and roundtables. When the Rio+20 process started, it was clear from the UNGA Resolution that governments wanted stakeholders to be even more engaged in the process than before. The Major Groups programme coordinator for the Rio+20 Secretariat was Chantal Line Carpentier – a true force of unlimited energy from Québec in Canada, who right through the Rio+20 process still had spare energy to continue one of her favourite leisure activities, taking part in Ironman races.[4]

During the years of the Commission on Sustainable Development, the United Nations had identified focal points for the nine Major Groups, which for the Rio+20 process became instrumental for outreaching to their respective communities and seeking stakeholder input to Rio+20. This was the first UN summit that had sought to integrate stakeholder views from the beginning of the preparatory process. This increased recognition was combined with a partnership agreement between the United Nations and Stakeholder Forum in order to see this NGO act as the complementary website to the United Nations for non-state stakeholders. In the process, Stakeholder Forum developed tremendous credibility with governments as an authority on sustainable development.

### *Box 2.5* Major Group organizing partners

From the very beginning of the preparatory process for the Earth Summit in 1992, people realized that sustainable development could not be achieved by governments alone. Agenda 21, the main outcome document from Rio, formalized nine sectors of society through which all citizens could participate in UN activities concerned with achieving sustainable development. These are officially called "Major Groups." This is a list of those who were the organizing partners for each of the Major Groups:

### *Business and industry*

- the International Chamber of Commerce: Louise Kantrow (www.icc.org);
- the International Council of Chemical Associations: Birgit Engelhardt (www.icca-chem.org);
- the World Business Council for Sustainable Development: Peter Paul van de Wijs (www.wbcsd.org/home.aspx);
- the UN Global Compact: Lila Karbassi (www.unglobalcompact.org).

### *Children and youth*

- Rio+Twenties (http://rioplustwenties.org), a youth-led, volunteer-based organization, served as the organizing partner for the MGCY, helping to create a platform for active youth participation in Rio+20: Sabá Loftus; Kiara Worth;
- the Belgrade Child Rights Centre: Ivana Savić.

### *Farmers*

- the World Farmers Organisation: Robynne Anderson (http://worldfarmersorganisation.com);
- La Via Campesina (International Peasant's Movement): Carlos Marentes and Jessica L. Roe (http://viacampesina.org/en/).

### *Indigenous peoples*

- Tebtebba (Indigenous Peoples' International Centre for Policy Research and Education): Vicky Tauli-Corpuz and Jocelyn Carino Nettleton (www.tebtebba.org);
- the Indigenous Environmental Network: Tom Goldtooth (www.ienearth.org).

### *Local authorities*

- ICLEI (Local Governments for Sustainability): Susanne Salz (www.iclei.org);

- United Cities and Local Governments (UCLG): Natalène Poisson (www.cities-localgovernments.org);
- the Network of Regional Governments for Sustainable Development (nrg4SD): Maruxa Díaz-Cardama García (www.nrg4sd.org).

### *Non-governmental organizations*

- CIVICUS (World Alliance for Citizen Participation): Jeffery Huffines and Henri Valot (www.civicus.org);
- ANPED (Northern Alliance for Sustainability): Leida Rijnhout (www.anped.org);
- Stakeholder Forum for a Sustainable Future: Jan-Gustav Strandenaes (www.earthsummit2012.org; www.stakeholderforum.org);
- Consumers International: Luis Flores (www.consumersinternational.org).

### *The scientific and technological community*

- the International Council for Science (ICSU): Peter Bates and Gisbert Glaser (www.icsu.org);
- the World Federation of Engineering Organizations (WFEO): Jorge Spitalnik (www.wfeo.net).

### *Women*

- Women in Europe for a Common Future (WECF): Sascha Gabizon (www.wecf);
- Voices of African Mothers: Sabina Anokye-Mensah and Nana-Fosu Randall (www.vamothers.org);
- the Women's Environment and Development Organization (WEDO): Eleanor Blomstrom (www.wedo.org).

### *Workers and trade unions*

- the International Trade Union Confederation (ITUC): Anabella Rosemberg (www.ituc-csi.org);
- Sustainlabour (International Labour Foundation for Sustainable Development): Laura Martin (http://assemblyrio20.sustainlabour.org/index.php?lang=EN; www.sustainlabour.org).

Rio+20 was also the first UN process for which stakeholders were encouraged to submit their input formally to the UN Secretariat even from the stage when the Zero Draft Outcome Document was being compiled. There were 493 submissions by Major Groups to the Rio+20 Zero Draft. Overall, there was broad uniformity between the different types of organization. A word cloud produced by Stakeholder Forum based on all submissions showed that Health/Well-being, Poverty Alleviation/Poverty Reduction/Poverty Eradication, Participation and Natural Resources/Resource Management appeared in the top four rankings for Major Groups, Political Groups, and Regional Preparatory Meetings. For the United Nations and intergovernmental organizations, participation ranked sixth, while Resilience appeared as fourth.

However, some notable differences included the following:

- Accountability. This term was seventh for Major Groups, but did not appear in the top ten for any other type of organization. It came sixteenth and twenty-first for the United Nations and intergovernmental organizations (IGOs) and member states respectively.
- National Sustainable Development Councils. This term was eighth for Regional Preparatory Meetings but did not make it to any other top ten. Member states had this term at forty-ninth. It is unclear why this discrepancy in interest existed.
- Consumption and Production. This term appeared seventh for member states and tenth for Political Groups, but only at seventeenth and eighteenth positions for Major Groups and the United Nations and IGOs respectively. This was a surprising finding (Stakeholder Forum, 2012).

In addition to their written contributions, stakeholders had a number of preparatory meetings of their own for Rio+20. The most significant ones have already been covered in this chapter. These were the UN DPI's sixty-fourth NGO Conference, "Sustainable Societies – Responsive Citizens"; the scientists' "Planet Under Pressure"; and the business and industry consultation in The Hague, where it became clear that on the issue of the push by the financial sector supported by NGOs and other stakeholders, the business and industry Major Group was split. The International Chamber of Commerce (ICC) was against, though the chair of its Environment Committee, Dow Chemicals, issued a press release in favour and there was strong support within the World Business Council on Sustainable Development (WBCSD) initially. Its attitude changed under pressure from some of its members.

In addition, stakeholders attended and gave input to all of the government-organized preparatory meetings, as well as to all formal and informal preparatory meetings. Stakeholders also set up a number of significant issue-based clusters and coalitions to influence the negotiations, mainly on governance, on agriculture, water, energy, corporate social responsibility and the green economy.

## *Governance: the Sustainable Governance 2012 Network*

The Sustainable Governance 2012 Network was a coalition managed by Stakeholder Forum and supported by the Climate and Sustainability Platform, the Northern Alliance for Sustainability (ANPED) and the International Union for the Conservation of Nature (IUCN). The network produced 19 think pieces on the subject of governance. The goal of the papers was to help governments become aware of options for better governance in relation to sustainable development to consider for Rio+20. Some of the ideas reflected in the papers that played out in Rio+20 were:

- reform of the UN Commission on Sustainable Development into a new Council on Sustainable Development (Strandaneas, 2011);
- reform of UNEP, including extending its membership to all countries (Ivanova, 2011);
- the setting up of a new inter-agency committee on sustainable development within the United Nations (Dodds, 2011);
- National Councils on Sustainable Development (Niestroy, 2012);
- a UN Office of High Commissioner for Future Generations (Horváth, 2012);
- ombudspersons for future generations (Göpel, 2012);
- enhancing science–policy links for global sustainability (Glaser and Bates, 2011);
- Principle 10 as a Convention (Banisar *et al.*, 2011).

Although not all these ideas were reflected in the final Rio+20 text, they were certainly debated by governments. The Network's approach of publishing papers and organizing events for governments to digest and debate the ideas helped contribute to a richer outcome for Rio+20.

## *The green economy: Global Transition 2012*

The core organizations of the grouping Global Transition 2012, in conjunction with Stakeholder Forum as convener, were the New Economics Foundation (NEF) and the New Economics Institute. The Green Economy Coalition and UNEP also collaborated closely in this grouping. The Global Transition Network's approach was similar to that of the Sustainable Governance 2012 Network, since it published 11 Challenge Papers addressing:

1. the green economy (Jenkins and Simms, 2011);
2. going beyond GDP (Seaford *et al.*, 2011);
3. global inequality (Woodward and Abdallah, 2011);
4. one planet living (Riddlestone, 2011);
5. the blue economy (Crilly, 2011);

6 energy resources and services (Johnson and Hälström, 2011);
7 food security (Nærstad, 2011);
8 managing natural capital (Greenfield, 2011);
9 global finance and banking (Zadek, 2011);
10 green jobs and skills (Rosemberg, 2011);
11 trade (Shrubsole, 2011).

A synthesis of the core policy ideas and policy recommendations of these papers was submitted to the Rio+20 Zero Draft consultation and also to the High-level Panel on Global Sustainability. A number of the key issues outlined in the Challenge Papers were reflected in the outputs of both processes. Perhaps the most influential paper was "Green Jobs and Skills," as it reflected much of the position of the trade union movement. By publishing these papers and other additional blogs and opinion pieces in the open forum of the "Global Transition 2012" website, this grouping helped build a broader coalition of support and understanding for what the trade unions were trying to accomplish. Over the 20 years from Rio 1992, the trade union movement has played an increasing role in the sustainable development negotiations, led in the first period by Lucien Royer and for Rio+20 by Anabella Rosemberg. Royer and Rosemberg have helped create a positive engagement with the environmental movement, which had often not appreciated that change means that people's jobs have to be reformed and that there needs to be a just and fair transition.

The papers "Managing Natural Capital" and "Beyond GDP" added to the growing call for the inclusion of nature within economic measures, although this was not a uniform view held among NGOs and indigenous peoples, who had deep reservations about the idea of costing nature. Their concerns were shared by some governments, the ALBA group in particular, the sentiments of which were succinctly framed in this way: "Putting a price on nature is not the solution and will only benefit big capital, while deepening the multiple crises we are facing" (Bolivian Climate Change Platform, 2012).

Another interesting and influential paper was by Simon Zadek, a former chief executive of the Institute of Social and Ethical Accountability, where he brought some new thinking regarding the agenda that should follow the financial crisis that began in 2008. This included reform of the sovereign wealth funds and credit rating agencies to include sustainability criteria. Even though his ideas were included within the report of the Global Sustainability Panel, they were not reflected in the Rio+20 Zero Draft. It is important to recognize that new thinking often takes a little while to be processed by governments.

Overall, through interactive dialogue based on expert thinking, the aims of the Global Transition 2012 Initiative of catalysing the debate on the fundamental aspects of the transition towards the new economy were galvanized in the lead-up to the Rio+20 processes.

### *Paragraph 47: Dialogue on a Convention for Corporate Social Responsibility and Accountability*

The Dialogue on a Convention for Corporate Social Responsibility and Accountability was a joint initiative of Stakeholder Forum and Vitae Civilis from Brazil that aimed to set up a dialogue at Rio+20 on the need for such a convention by engaging representatives of the North and South. The approach taken was again to use background papers and workshops as a mechanism through which to engage governments in the idea. This was the third attempt by NGOs at trying to secure a convention in this field. Similar attempts had been made in 1992 and 2002, but both had failed. The first dialogues organized in New York discussed two proposals for such a convention. The first of these was a comprehensive framework addressing the liability of businesses for their environmental and social impacts, transparency and disclosure, human rights, relationship with communities and other stakeholders, and supply chains responsibility. The second proposal was a convention on corporate sustainability reporting, a framework that would mandate governments to implement policies requiring large companies to develop and disclose sustainability information in their annual reports.

Throughout the Rio+20 negotiation rounds, it was clear that governments were reluctant to discuss broader corporate social responsibility issues, and the discussion quickly narrowed to sustainability reporting policies, a topic already mentioned in the Rio+20 Zero Draft in January 2012. Therefore, efforts focused on protecting the text already in the negotiating document around having an international framework on sustainability reporting and moving from voluntary commitments to government regulation. Furthermore, the difference this time was that the approach was led by several key financial institutions. Aviva, the largest UK insurer and one of the top ten in the world, convened a "Coalition on Corporate Sustainability Reporting" and called for a Convention on Corporate Sustainability Reporting. It was joined by other significant investors, including Co-operative Asset Management, Calvert Investment, BT Pension Scheme Management Ltd and the Church Commissioners for England, which amounted to a combined portfolio of over $US2 trillion. Aviva (2011) identified two core elements for such a convention:

> First, the Convention would be a commitment by UN member states to develop national regulations, formal codes or listing rules that encourage the integration of material sustainability issues within the annual report of all listed and large private companies. Second, in order to be flexible, it would establish an opt-out for those companies that elect not to prepare such a report: they would be required to explain their rationale to their shareholders, creditors and other stakeholders. In other words, corporate sustainability disclosure would be on a "report or explain" basis.

A mandatory sustainability reporting agenda was supported by a diversity of groups. It was part of the key asks of a group of leading NGOs and other stakeholders (including Vitae Civilis), which called for the strengthening of corporate reporting on social and environmental impacts. Likewise, the IUCN and the World Business Council for Sustainable Development issued a joint statement in support of strengthening the "Future We Want" paragraph on corporate sustainability reporting. Organizers and partners brought together and amplified these messages through their dialogue events and awareness-raising activities during the negotiation rounds.

An important element to this work was bridging the North–South divide. The Dialogue, in coordination with the Corporate Sustainability Reporting Coalition, disseminated relevant information to strengthen the case for implementing sustainability reporting policies. A number of governments were already doing what was being suggested. For example, the South African stock exchange had, through the recommendations of the King III report (King, 2009), mandated listed companies to disclose their sustainability information and produce reports. The São Paulo stock exchange was strongly encouraging its listed companies to produce their sustainability reports on a comply or explain basis. At the end of the Rio+20 process, although the outcome was considerably weaker than expected, governments agreed to maintain the discussions on how corporate sustainability reporting policies could be expanded through the identification of best practices and the participation of stakeholders. A group of governments called Friends of Paragraph 47 (Brazil, Denmark, France and South Africa) was formed shortly after Rio+20 to lead the discussions.

### *Rio Principle 10: access to information, public participation and environmental justice*

Principle 10 of the Rio Declaration pertains to access to information, public participation and environmental justice. Since the 1992 Rio Conference, the UN Economic Commission for Europe (UNECE) has developed Principle 10 into a convention known as the Aarhus Convention that came into force on October 30 2001. Around the 2002 Johannesburg Summit there was discussion about trying to extend Aarhus into a global convention or to make it the basis for developing a global convention. This was greeted in 2002 by many developing countries as an attempt to impose a European convention or a European approach on them. This attempt informed the approach taken by World Resources Institute (WRI) and the other members of the "Access Initiative." The Access Initiative initially supported the idea of developing other regional conventions with a goal of joining them together through a framework convention. In fact, there seemed to be interest in Latin America in developing a regional convention. Lalanath de Silva of the WRI reminded us that "[s]ince 1992, Principle 10 has spawned laws providing access to environmental information in more than 100 countries, public participation provisions in more than 120 countries, and environmental courts and tribunals in more than 44 countries" (quoted in Dodds and Nayar, 2012).

By the UN DPI NGO conference in September 2011, there was a strategic disagreement between the NGOs. The European Environment Bureau – the largest environment coalition in Europe, now headed by Jeremy Wates, former executive secretary of the Aarhus Convention – wanted, and believed it was possible to still get, a commitment for a global convention at Rio+20. Ultimately, NGOs decided to support the call for the global convention because it did not rule out the possibility of regional conventions if a global one proved a step too far. This view was included in paragraph 58 of the Zero Draft of "The Future We Want": "agree to take steps to give further effect to Rio Principle 10 at the global, regional and national level, as appropriate" (United Nations, 2012b). "Although Rio+20 did not give rise to a global convention on Principle 10, paragraph 99 [of "The Future We Want"] 'expressly encourages action at the regional level', which has opened the possibility to negotiate regional conventions on Principle 10" (Dodds and Nayar, 2012). Soon after the outcome text was revealed, Ambassador José Luis Balmaceda of Chile, joined by a number of Latin American countries, including Mexico, called for a Latin American and Caribbean Convention on Principle 10 at the "Choosing Our Future" event in Rio de Janeiro (ibid.). Since then, 11 countries in Latin America have joined the government of Chile and announced their intent to go forward with the second regional instrument on Principle 10 by following the lead Europe took with the Aarhus Convention.

## The transition from the informal and the global to the textual of "The Future We Want"

On January 10, 2012, all permanent representatives to the United Nations received a letter jointly signed by Ambassador John W. Ashe of Antigua and Barbuda and Ambassador Kim Sook of Korea, co-chairs of the Rio+20 Preparatory Committee, in which they distributed the "Zero Draft" of the Outcome Document of Rio+20, which was meant to be the basis for negotiation during the initial discussions scheduled for January 25–27, 2012.

The letter also gave guidance on logistical matters, confirming three days of meetings, with the initial three sessions of those days being dedicated to "informal-informal" negotiations on sections I and II of the Zero Draft. Member states were requested to submit written comments on sections I and II by January 23 and on sections III, IV and V by February 17. The letter concluded with the co-chairs expressing appreciation to the UN Secretary-General for his support, hoping that the report of his Global Sustainability Panel would be available shortly and extending good wishes to recipients for the year 2012. A private portal was established via which member states, the United Nations' nine official Major Groups and registered NGOs were able to submit comments electronically. The Zero Draft, which was also dated January 10, 2012, comprised 128 paragraphs over 19 pages and was divided into five sections:

- section I, Preamble and Vision, paragraphs 1–5;
- section II, Renewing Political Commitment, paragraphs 6–24;
- section III, Green Economy in the Context of Sustainable Development and Poverty Eradication, paragraphs 25–43;
- section IV, Institutional Framework for Sustainable Development, paragraphs 44–62;
- section V, Framework for Action and Follow-up, paragraphs 63–128.

Fifteen days later, at 10.00 a.m. on January 25, ambassador and co-chair John W. Ashe started the Rio+20 negotiating process with the words "I call to order this informal, initial discussion on the zero draft of the outcome document for the United Nations Commission on Sustainable Development" (Ashe, 2012). Ambassador Ashe's statement was followed by one by Sha Zukang, the secretary-general of the conference (Sha Zukang, 2012).

And thereafter thus began the onslaught of criticism of what had been a keenly anticipated document. Member states and UN groups made critical comments that would be repeated throughout the preparatory process and the negotiations, and even after negotiations had finished and the Outcome Document had been finalized. The criticism nearly led to an all-out rejection of the newborn by its family.

The comments made by the various political groups were scathing. The African Group asserted that the Zero Draft lacked balance and failed to address some of the critical issues of the African region. The Group commented that "the zero draft lacked vision and ambition . . . much work is needed to turn it into a bold vision to achieve sustainable development based on three paramount objectives, implementation, integration and coherence." The Association of Small Island States (AOSIS) commented in similar terms and expressed concern about the document's "insufficient attention" to SIDS issues and spoke of a shared responsibility to radically improve it and quickly: "all nations must work together to raise our aspirations to deliver a more ambitious document." The European Union, while grateful for the efforts that had been made to compile thousands of pages of inputs in a streamlined document, concluded that the draft could "serve as a useful common basis for starting the negotiations . . . but a number of critical issues have not been adequately addressed at all." The Union indicated that it was "concerned about balance, and consistency in the text and the lack of ambition in certain domains." The Group of 77 saw the document as being devoid of "vision, ambition, balance and action-oriented language." The LDCs called for "the incorporation in the Zero Draft of stronger commitments for action to address the particular needs and concerns of the LDCs."

The Major Groups (business and industry, children and youth, farmers, indigenous peoples, local authorities, NGOs, science and technology, women, and workers and trade unions), all commented. While they were generally supportive and praised aspects of the document, they all expressed concern about its "lack of ambition," which they unanimously felt was urgent. They also stressed the need for greater emphasis on their respective areas of interest. The expression "lack of ambition" would stick in the public dialogue and perception, and would in the end also be applied to "The Future We Want" as the Outcome Document of the Rio+20 Conference.

Against this torrent of criticism it is necessary to understand the background and contents of the Zero Draft. The General Assembly Resolution A/RES/64/236 of December 24, 2009 approved the hosting of the Rio+20 Conference and mandated that the outcome should be "a focused political document." In producing a Zero Draft of 20 pages, the drafters were trying to be faithful to that mandate. The drafters also ensured that the seven priority areas as identified in the submissions to the compilation document (United Nations, 2012e) were addressed in the Zero Draft. These seven priority areas (United Nations, 2011b) were "decent jobs, energy, sustainable cities, food security and sustainable agriculture, water, oceans and disaster readiness." Second, the Zero Draft was the product of a member state-led process that started in 2010 with the dissemination of a questionnaire to member states by UN DESA, to which they had been urged to respond, sharing views on the themes and objectives of Rio+20 and their experiences on the subject areas of the conference.

A mere 25 per cent of member states responded to the questionnaire, and those responses formed the basis for a synthesis report or compilation document. According to the CSD, "[t]his compilation document is to serve as the basis for the preparation of a zero-draft of the outcome document, to be presented for consideration by Member States and other stakeholders by January 2012" (United Nations, 2011c). For all the criticisms member states made of the Zero Draft, it was a reflection of what the member states themselves had actually offered. In addition to the comments on the Compilation Document, further comment was invited on the Zero Draft, and UN DESA received some 677 responses and comments, which caused the Zero Draft to swell from a "focused political document" to one that sought to capture every conceivable issue.

The Zero Draft of "The Future We Want" contained the core of all the elements that would come to distinguish Rio+20. It raised the issue of planetary boundaries, firmly linked poverty eradication to the green economy, a concept that was raised for the first time in a multilateral agreement, and spoke for an enhanced role for civil society. It reaffirmed the central role of the General Assembly, of the High-Level Political Forum as an instrument for dealing with issues of sustainable development, and of ECOSOC in addressing these, as well as social and economic, matters. It also spoke to a more effective Commission on Sustainable Development and for UNEP to become a strengthened and specialized agency with universal membership. It raised the proposal of an Ombudsman for Future Generations and included the controversial issue of consumption and production. Unlike the final agreed declaration, the Zero Draft addressed the issue of green economy toolkits. The Zero Draft raised for the first time in a multilateral agreement the prospect of a zero land degradation world, or, in the language used, a "land degradation neutral world."

The broader topics with which the Zero Draft dealt were water, energy (including endorsing and recognizing the benefits of SE4All), cities, green jobs and social inclusion, oceans, SIDS, LDCs, natural disasters, climate change, forests and biodiversity, land degradation and desertification, mountains, chemicals and wastes, education, and advancing women's rights and greater economic prosperity for women (although it did not deal with the issue of women's reproductive rights, which became one of the strongest criticisms made against the final Outcome Document). The Zero Draft also introduced the possibility of a list of SDGs to replace the MDGs, dealt with means of implementation, including financing for development, science and technology, capacity-building and trade. It advanced the idea of a knowledge-sharing platform and the transformative benefits of green investments, and proposed a Compendium or Registry of Commitments, which in attracting over 700 commitments at the closing of Rio+20 proved to be highly successful.

After member states, Major Groups and other stakeholders commented on the Zero Draft, it grew from its original 19 pages to a colossal and unwieldy 200 pages. The draft Rio+20 Outcome Document was eventually pared down to 50 pages and an attempt was made to try to realign it with the mandate of the General Assembly Resolution. All in all, in retrospect it can be said that the original Zero Draft was a sound document and formed the foundation on which the Rio+20 Outcome Document was built, which, despite criticisms, as will be discussed later, contained seminal recommendations that were unique to the multilateral development dialogue and agenda.

## Enabling UN leaders behind Rio+20

Aside from their responsibilities as high serving officers, many leaders in the UN system heeded the call of the UN Secretary-General and developed a holistic view towards sustainable development. For instance, the phenomenal work of UN Volunteers attracted commitments to various aspects of volunteerism by 64 million people.

### *Sha Zukang*

Sha Zukang was the secretary-general of Rio+20. A career diplomat, he became the UN Under-Secretary-General for Economic and Social Affairs on July 12, 2007 and headed the Department of Economic and Social Affairs (DESA). Previously he had been the Permanent Representative of the Chinese mission to the UN office in Geneva and chair both of the Preparatory Committee and of the Committee of the Whole for the United Nations Conference on Trade and Development (UNCTAD) eleventh session in 2003 and 2004. He had also been president of the Trade and Development Board for the fiftieth session of

*Figure 2.5* Despite great interest in Rio+20 by youth, it proved impossible to agree on a figure that represents future generations. The image shows a demonstration outside the meeting rooms during the final days of negotiations (courtesy of IISD/*Earth Negotiations Bulletin* (www.iisd.ca)).

UNCTAD, chair of the Government group of the Governing Body of the International Labour Organization in 2002 and 2003, and a member of the UN Secretary-General's Advisory Board on Disarmament Matters from 1994 to 1999. In addition, he has served as president, vice-president, chair, coordinator and expert in many international conferences in the field of arms control, trade, intellectual property, social affairs and telecommunications, among others.

He established the Department of Arms Control of the Ministry of Foreign Affairs of China and became its first director-general. He participated, as chief negotiator or representative of the Chinese government, in the negotiation and review of many important international treaties on arms control and disarmament, such as the Treaty on the Non-Proliferation of Nuclear Weapons, the Comprehensive Nuclear Test Ban Treaty, the Chemical Weapons Convention, the Biological and Toxin Weapons Convention and the Convention on Certain Conventional Weapons. He made great contributions to mine clearance in China and many parts of the world. He encouraged the development of Chinese non-governmental organizations, and facilitated the opening of offices in China by international organizations.

As secretary-general of the conference, Sha Zukang ensured that Rio+20 would not be qualified as a non-transparent and opaque process. From the onset, he, with the support of Ambassador John W. Ashe, made it clear that they were determined to have a transparent, inclusive process, open to all the parties and with equitable treatment for all stakeholders. Sha Zukang repeatedly stressed and sought to reassure member states that the process was to be driven by them. His opening remarks at the first informal consultation on the Zero Draft, which took place on January 25, 2012, reflect his vision as leader of the conference: "We, in the Secretariat, will do everything in our power to support you: your commitment your ambition and your action. And we look to you to guide us" (Sha Zukang, 2012). By repeating such sentiments throughout the duration of the process, Sha Zukang sought to build confidence in the transparency and equity of the process by making it clear that the Rio+20 Secretariat would shepherd the process but not hijack it, that member states owned and were responsible for the process, the negotiations, the outcome and the final text. In advancing the process of transparency and paper saving, UN DESA opted for posting all Rio+20 materials on a dedicated website, including questionnaires and synthesis reports, the Compilation Document and the amendments proposed for the Zero Draft. This was a new and important step in shining the light of transparency into the UN process, in allowing all stakeholders equal and simultaneous access to what was being contributed, indicating how and by whom it was incorporated. There was a section for member states themselves, which contained documents that were particularly relevant to countries. In this way, UN DESA, acting on behalf of the UN system, provided a broader level of access, increased the means of communication and effectively enhanced the democratic process.

### *Elizabeth Thompson*

Elizabeth Thompson was appointed by the secretary-general of the United Nations as one of two executive coordinators for the UNCSD Rio+20 Conference (one coordinator was drawn from the geopolitical North and one from the South). She assumed her duties on December 7, 2010, reporting directly to Sha Zukang. Thompson had been involved in negotiations on climate change and sustainable development since 1994. She was a government minister in Barbados and an elected parliamentarian from 1994 to 2008, serving in several ministries over that period, including as Minister of Energy and the Environment, Minister for Physical Development and Planning, Minister of Housing and Minister for Health. In 2008 she was awarded UNEP's Champion of the Earth award.

### *Brice Lalonde*

Brice Lalonde was appointed by the Secretary-General of the United Nations as the second executive coordinator for Rio+20, taking office on January 1, 2011. He reported directly to Sha Zukang. Lalonde was a former French environmental activist and Minister of the Environment who had deep experience working across the global environmental agenda. At the time of his appointment to the Rio+20 Secretariat, he was serving as France's climate change ambassador, working in the UN Framework Convention on Climate Change negotiation process. He remained deeply committed to the French and EU positions that advocated for a new international environmental organization as a specific outcome of the Rio+20 Conference.

### *Nikhil Seth*

Nikhil Seth took over the Rio+20 Secretariat as director of the UN Division for Sustainable Development within UN DESA in late 2011, following Tariq Banuri's retirement. He has worked in the United Nations since 1992 and has served as special assistant and chief of office to the Under-Secretary-General for Economic and Social Affairs, and as secretary of the Economic and Social Council (ECOSOC) and the Second Committee of the General Assembly. Prior to taking over the Secretariat for Rio+20, Seth was director of the UN DESA Office for ECOSOC Support and Coordination (OESC), where he guided ECOSOC's work on implementing several new key mandates, including its Annual Ministerial Review (AMR) and Development Cooperation Forum (DCF). In 1992 he was a member of the Indian government's delegation to the Rio Earth Summit.

Nikhil Seth played a vital role in ensuring the coherence of the Secretariat support for Rio+20. He also sought to make available to government representatives the necessary information on the new ideas that were being introduced, in particular the green economy and the SDGs. He built on the support that his predecessor had given to stakeholders, and this helped create greater awareness concerning, and outreach for, the conference. He also helped in building acceptance among member states of what the conference might achieve, which at times was not easy. One of the principal contributions that he made was to ensure the UN system worked together in task teams and produced thought pieces and issues briefs that would allow for valuable dialogue and heighten constructive engagement amongst stakeholders on the themes and objectives of the conference. This was of critical importance, given the fact that concepts which did not yet exist in the global multilateral agenda were being introduced, such as the green economy, natural capital accounting, a sustainability index, land degradation neutrality, valuation of ecosystem services, and the SDGs. It was also important in helping to obtain agreement on the ten-year framework for sustainable consumption and production.

### *Tariq Banuri*

Tariq Banuri served as director of the UN Division for Sustainable Development in UN DESA and the Rio+20 Secretariat until his retirement in the autumn of 2011. He had been a strong advocate for sustainable energy, playing a critical role in the conference, and reopened the door for a much more collaborative relationship with stakeholders. Earlier, he had been instrumental in the design of a number of institutions and networks on sustainable development, including the Sustainable Development Policy Institute (SDPI), the Sarhad Rural Support Programme (SRSP), the Asia Centre of the Stockholm Environment Institute (SEI) and the United Nations Office for Sustainable Development in Inchon, South Korea. Tariq Banuri was also a coordinating lead author on the Intergovernmental Panel on Climate Change (IPCC), which received the Nobel Peace Prize in 2007. After retiring from the United Nations he joined the University of Utah's Department of City and Metropolitan Planning.

### *David O'Connor*

David O'Connor's role in Rio+20, although played largely behind the scenes, was invaluable. His formal job title is Chief of the Policy and Analysis Branch of UN DESA's Division for Sustainable Development. In that capacity he had responsibility for leading UN DESA's Rio+20 team in policy formulation, the development of thought leadership and papers for the conference. This included drafting, disseminating and collating the responses to the questionnaire that UN DESA distributed to member states and stakeholders, the answers to which formed the basis of the Synthesis Report, and SD21 ("Sustainable Development in the 21st Century"). His responsibilities also encompassed drafting the co-chairs' Zero Draft, as well as the Herculean task of reducing the Compilation Document of 6,000 pages – comprising 677 submissions from 100 member states, 493 from NGOs and Major Groups, 74 from UN agencies and IGOs, 4 from political groups and the outcomes of 5 regional meetings – into a concise, accurate, accessible text, while still capturing the essence of their myriad and diverse positions.

### *Janos Pasztor*

Janos Pasztor was a highly experienced international civil servant who served as the executive secretary of the UN Secretary-General's High-Level Panel on Global Sustainability (GSP). He shepherded the production of the GSP's well-received report *Resilient People, Resilient Planet: A Future Worth Choosing*. He brought a rich and diverse professional background and experience to the post, encompassing energy, environment and climate change, and work in the NGO community and within the UN system. His service to the GSP followed roles as director of the UN Secretary-General's Climate Support Team, in UNEP and as director of the UN Environmental Management Group. Months after Rio+20, he joined WWF International as Director of Policy and Science.

### *Achim Steiner*

In 2006, Achim Steiner was appointed UN Under-Secretary-General and executive director of UNEP. Steiner was one of the first leaders to embrace with visionary perspective the idea of a conference on sustainable development. In doing so, he planted many of the seeds that would evolve into the transformation of a convergent sustainable development agenda, which joins the heritage of the Millennium Declaration and of the Rio principles on environment and development. Starting in 2007, Steiner made sure that UNEP, with the endorsement of the members of its Governing Council, would be able to provide the scientific arguments needed to justify the convergence between the planet-centred and the people-centred development agendas. *Global Environmental Outlook 4* – the most comprehensive global environmental assessment – advocated for a simple yet powerful concept: "environment for development."

In October 2008, in the wake of the economic and financial crisis, Steiner was the first proponent within the United Nations for a "global green new deal." This initiative, inspired by Franklin D. Roosevelt's recovery-focused "New Deal," provided the first coherent analytical framework for the economic policy instruments that would later form the basis of the green economy toolkits and platforms contained in Rio+20's "The Future We Want." The idea behind the global green new deal was to catalyse action and strong political engagement to direct financial flows and manage markets to deal with the even greater global challenges of our time, thus generating businesses in renewable energies, clean technology ventures, sustainable agriculture, conservation and the intelligent management of the planet's ecosystems and nature-based infrastructure. As a continuation of this initiative, and upon Steiner's inspiration, the UN agencies, funds and programmes launched a Knowledge Project entitled "Towards a Green Economy: Pathways to Sustainable Development and Poverty Eradication."

Steiner did not have any formal role, but UNEP also benefited from the Rio+20 theme of the Institutional Framework for Sustainable Development and the decision of the conference to strengthen and upgrade UNEP. However, it was not agreed that UNEP should be converted to a full UN agency to be known as the World Environment Organization or the United Nations Environment Organization, as UNEP and some countries, especially European ones, had pressed for.

Ahead of Rio+20, UNEP produced substantive and critical reports (United Nations Environment Programme, 2011) on the green economy: the seminal but also controversial (to some countries and stakeholders) *The Economics of Ecosystems and Biodiversity* (*TEEB*), and *GEO-5* on the state of the planet – that contributed significantly to the preparatory and post-Rio+20 processes and created important source documents for up-to-date scientific and policy data. *TEEB* was a ground-breaking work that UNEP introduced into the dialogue. It was "an international initiative to draw attention to and highlight the growing cost of biodiversity loss, ecosystem degradation and to help frame an objective global standard for Natural Capital accounting." It drew attention to the role and value of ecosystems services global in society and economy.

### *Kandeh Yumekella*

Kandeh Yumekella, director-general of the United Nations Industrial Organization (UNIDO), played a significant role in Rio+20. In June 2009 he was appointed by UN Secretary-General Ban Ki-moon to co-chair a new Advisory Group on Sustainable Energy for All. He also chaired UN Energy, the inter-agency group which also input substance from the UN system on the issue of energy. UNIDO and the UNDP collaborated on the hosting of the SIDS launch of SE4All, which was highly successful and the only regional preparatory meeting at which member states endorsed the SE4All initiative.

### *Rachel Kyte*

Rachel Kyte is vice-president of sustainable development at the World Bank. She had responsibility for the Bank's input to Rio+20, driving the World Bank's leadership on inclusive green growth and climate change. Prior to her appointment, she was vice-president for business advisory services at the International Finance Corporation. In 1992 she worked for the Centre for Our Common Future, which organized the Rio Global Forum around the Earth Summit. She went on to be executive director of the Women's Environment and Development Organization after Rio 1992.

Although the World Bank is not a member of the UN system family, its and Rachel Kyte's contribution to Rio+20 should not be underestimated. She played a fundamental role in ensuring the World Bank's support in the policy discussions on the green economy by providing constant examples of how adequate management and accounting of a country's natural assets transformed its developmental perspectives. She also contributed towards a constructive policy discussion on measuring progress beyond GDP and in laying the ground for the Bank's support for the SDGs as part of an integrated development agenda beyond 2015.

### *Alicia Bárcena*

Alicia Bárcena, under secretary-general at the Economic Commission for Latin America and the Caribbean (ECLAC), had been part of the original Rio Secretariat in 1992, where she had been in charge of stakeholder engagement. She moved on to be the first executive director of the Earth Council, which had been set up by Maurice Strong to take forward the outcomes of Rio 1992 and hold governments accountable. In 1997 she rejoined the United Nations, and for Rio+20 she headed ECLAC, one of the most active UN Regional Commissions, hosted meetings and produced valuable reports relating to Latin America and the Caribbean and SIDS. Bárcena and ECLAC's contribution to Rio+20 helped in drawing global attention to the structural linkages between inequalities, multidimensional poverty and sustainable development. She advocated for a development agenda that encourages developing countries, particularly the majority of middle income countries, to move beyond an agenda based on the fulfillment of basic human needs to a more ambitious one aiming at bridging persistent inequalities at the national, regional and global levels.

### Olav Kjørven

The UNDP was late joining in the Rio+20 process, only really engaging after the submission date for the Zero Draft. The UNDP supported national dialogues and preparatory meetings in some member states. As Rio+20 approached, Olav Kjørven, UNDP assistant secretary-general and director of the Bureau for Development Policy, took charge of the UNDP's growing Rio+20 team. One of the outputs of Rio+20 was the idea that in many respects GDP, with its exclusive focus on economic data, excludes the full gamut of social and environmental factors that contribute to development. The possibility of the formulation of a Social Development Index or a set of indicators was discussed at Rio+20 and included in the text, although no formula was agreed on for such an index. At the end of the Rio+20 Conference, the UNDP, which currently has responsibility for the production and publication of the globally used Human Development Index (HDI), proposed a template for a new sustainable development index called a "Human Sustainability Index" with a "conceptual framework that would recognize rates of human development while also weighing up the cost of progress to future generations." Drawing from the experiences of Rio+20 Kjørven helped to mobilize the UN system to start the global consultations for preparing the Post-2015 Development Agenda.

### Veerle Vandeweerd

Veerle Vandeweerd is director of Environment and Energy at the UNDP, providing strategic direction and oversight to the programme in mainstreaming environment, energy and sustainability into its development work in 166 countries. Vandeweerd oversees the delivery of an extensive portfolio of projects and funds, including the UNDP's Global Environment Facility and the Montreal Protocol Unit. The energy and climate change portfolio covers the full policy cycle, from project development to monitoring, reporting and verification (MRV), and includes all thematic areas ranging from Low Emissions Climate Resilient Development Strategies to Transformative Ecosystem Management. Before joining the UNDP in 2007, Vandeweerd held leadership roles within UNEP for 19 years, based in Africa and Europe. During the whole Rio+20 process from 2009, she was a major and consistent advocate for the active involvement of the UNDP in the proposed Rio+20 Conference.

## The Rio+20 Secretariat

The United Nations Department for Economic and Social Affairs (UN DESA) was responsible for leading the contribution of the UN system, as well as for the organization and leadership of the conference. UN DESA's contributions to the organization of the conference, its logistical and substantive arrangements, the negotiations and the Outcome Document are noteworthy.

In order to ensure that ideas for discussion were made available to all stakeholders, the UN system had to create thought pieces around some of these new ideas. The system did not fall short of its responsibility, and a plethora of papers was produced (UN DESA, 2012a) to support stakeholder cerebration and consultations. The United Nations must be given recognition for helping to guide the process by making available, to all stakeholders and the wider public, briefs and papers that would stimulate discussion and help crystallize the thought and decision-making processes of member states. The UN system's production of papers served to provide clarity and disseminate information on complex issues. This helped to prevent the process from becoming derailed, and contributed to a consensus outcome. The list that follows (UN DESA, 2012b) is not comprehensive but gives some of the topics and titles of briefs and papers that were produced for the conference:

- cities;
- disaster reduction;
- finance;
- food security and agriculture;
- institutional framework for sustainable development (IFSD);
- *GEO-5*: environment for the future we want;
- governance;
- green economy in SIDS: an analysis of challenges and opportunities;
- green economy report, *Towards a Green Economy*;
- green growth and climate resilience national strategies for climate change and low carbon;
- green jobs;
- migration;
- migration and sustainable development;
- oceans;
- population;
- population dynamics and sustainable development;
- science and technology;
- the Secretary-General's Report on Objectives and Themes of the UNCSD;
- Sustainable Development Goals and indicators;

- sustainable development in Latin America and the Caribbean: 20 years on from the Earth Summit: progress, gaps and strategic guidelines;
- sustainable tourism;
- *TEEB*;
- trade and the green economy;
- transport;
- water.

On a "paper smart" move, the United Nations stated that "in a move towards greater sustainability, Rio+20 is the first major UN conference to carry out a 'PaperSmart' programme to cut paper use, vastly reducing paper consumption by close to 20 million sheets" (UN DESA, 2012c). This was another of Rio+20's unique accomplishments. In cutting the paper usage in the preparatory process by accepting online submissions and then in making the conference "paper smart," the UN system saved millions of reams of paper and their source trees, and added a different dimension of leadership to the conference.

### *The contribution of the Global Compact*

The Global Compact made a significant contribution to Rio+20. At its Corporate Sustainability Forum (CSF, 2012), held in Rio de Janeiro during the Third Preparatory Committee, the Global Compact brought together some 3,000 businesspersons in a heightened level of engagement, and record attendance for the private sector at a multilateral governmental conference. The Compact's Business Forum will be explored in more detail in Chapter 4, but it can be said here that it was one of the accomplishments of the Rio+20 process and conference. The Forum provided businesspeople with the opportunity to consider what enabling environment they need for a transition to a global green economy; how shareholder value can be driven in the new resource-constrained world; how green business practices can define and give competitive advantage to brands; and why practising sustainability is in their own self-interest. It was the aim of the Forum to "strengthen the business contribution to sustainable development globally – seeking to bring greater scale to responsible business practices, to advance and diffuse sustainable innovation, and to stimulate broader collaboration between companies, governments, civil society and the UN" (ibid.).

Education is central to creating awareness, disseminating information and knowledge, and changing behaviours. It is therefore an essential tool for building a culture of sustainability and fostering sustainable development globally. With this recognition, the Global Compact-Principles for Responsible Management Education (PRME) gave priority to the launch at Rio+20 of the Higher Education Sustainability Initiative (HESI) and to hosting a dialogue on how education can advance sustainability. The HESI, which will be discussed in Chapter 4, is yet another area in which the invaluable role played by the UN system can be recognized. In promoting its education efforts, PRME (2012) noted that

> [b]usiness schools, management-related academic institutions, and universities have a unique role to train current and future generations to lead this process. However, as a global sector, management education must make considerable change to be at the forefront of innovation and progress for sustainable development.

### *The Department of Public Information*

The UN system also had responsibility, through its Department of Public Information (DPI), for public relations and messaging. The creativity and effectiveness of the DPI's public outreach and messaging for the Rio+20 Conference are highly debatable. Prior to the conference, it was repeatedly said that ambition was low and that it had failed to create a sense of excitement. However, in so far as there was an absence of excitement, this was as much the result of prevailing international conditions as it was of an underwhelming messaging and public relations effort by the DPI. These observations having been made, however, it must be conceded that with Felix Dodds chairing the meeting and leading the work on the outcome text, the DPI convened the very successful sixty-fourth UN DPI NGO Conference (United Nations, 2011b), the recommendations from which later informed the negotiating process and the Rio+20 Outcome Document. The DPI-NGO Conference will be considered later but it helped to catalyse support for Rio+20 in the NGO and other stakeholders communities, to such an extent that in May 2012, Rio+20 attracted the largest ever gathering of people at a UN meeting, with a massive NGO presence.

## Notes

1 Leida Rijnhout (ANPED), Juan Hoffmaister (TWN), Jan-Gustav Strandenaes (ANPED), Uchita de Zoysa (The Widening Circle), Aron Belinky (Brazilian Civil Society Facilitating Committee), Jeremy Wates (European Environmental Bureau), Lalanath de Silva (WRI), Neth Daño (ETC Group), Nilab Alokuzay (BIMUN), Anne Sturm (Association Internationale des Charités), Suzanne Salt (ICLEI) and Mirian Vilela (Earth Charter International).

2 Colombia shared with the authors a copy of the meeting's notes, which they took and kept for their internal use.

3 Information about REN21 can be found online at: www.ren21.net

4 An Ironman Triathlon is one of a series of long-distance triathlon races organized by the World Triathlon Corporation (WTC) consisting of a 2.4-mile (3.86-kilometre) swim, a 112-mile (180.25-kilometre) bicycle ride and a marathon 26.2-mile (42.2-kilometre) run, raced in that order and without a break. For more information, see Wikipedia: http://en.wikipedia.org/wiki/Ironman_Triathlon

## References

Ashe, J. W. (2012) Opening comments at first PrepCom. Available online at: https://docs.google.com/gview?url=http://www.uncsd2012.org//content/documents/651ashe.pdf&chrome=true

Australian, The (2010) I Never Liked You, Chinese Diplomat tells UN Secretary General Ban Ki-moon in Drunken Rant, *The Australian* (Sydney), September 9. Available online at: www.theaustralian.com.au/news/world/i-never-liked-you-chinese-diplomat-tells-un-secretary-general-ban-ki-moon-in-drunken-rant/story-e6frg6so-1225916507021

AVIVA (2011) Earth Summit 2012: Towards a Convention on Corporate Sustainability Reporting at Rio+20. Available online at https://www.aviva.com/data/media-uploads/news/File/Rio%202012/Towards%20a%20convention%20on%20corporate%20sustainability%20reporting%20two%20pager%20end%20March%202012.pdf

Banisar, D., Excell, C. and de Silva, L. (2011) Moving from Principles to Rights: Rio 2012 and Ensuring Access to Information, Public Participation, and Access to Justice for Everyone, London, Stakeholder Forum. Available online at: www.stakeholderforum.org/fileadmin/files/Moving%20from%20Principles%20to%20Rights.pdf

Black, R. (2012) Rio Revisited: Glass Half-Full?, BBC News, July 24. Available online at: www.bbc.co.uk/news/science-environment-18967011

Bolivian Climate Change Platform (2012) Position of the Bolivian Climate Change Platform on Rio+20 and the Green Economy, La Paz, Bolivian Climate Change Platform. Available online at: www.cambioclimatico.org.bo/derechosmt/052012/100512_2.pdf

Bonn 2011 Nexus Conference (2011) Reflections of the Co-chairs on the Outcomes of the Bonn 2011 Nexus-Conference. Available online at: www.water-energy-food.org/en/conference/reflections.html

Colombia (2011) Rio+20: Sustainable Development Goals (SDGs) Concept Note. Available online at: www.uncsd2012.org/content/documents/colombiasdgs.pdf

Colombia, Guatemala and Peru (2011) Rio+20: Sustainable Development Goals: An Updated Proposal, November. Available online at: www.uncsd2012.org/content/documents/539Update%20of%20the%20SDG%20Proposal%20ENG%20DEC11.pdf

Colombia, Peru and the United Arab Emirates (2011) Concept Note on Sustainable Development Goals. Available online at: www.minambiente.gov.co/documentos/DocumentosInstitucional/rio_20/150512_proposal_colombia_emirates.pdf

Commission on Climate Change and Development (2009) Closing the Gaps, Stockholm, Swedish Ministry of Foreign Affairs. Available online at: www.ccdcommission.org/Filer/report/CCD_REPORT.pdf

Crilly, R. (2011) The Blue Economy: Fishing Public Resources for Public Benefit, Global Transition 2012, London. Available online at: http://stakeholderforum.org/sf/outreach/index.php/int3day2home/82-inf1day2/660-ifn1day2item4

CSF (2012) Rio+20 Corporate Sustainability Forum. Available online at: http://csf.compact4rio.org/events/rio-20-corporate-sustainability-forum/event-summary-251b87a2deaa4e56a3e00ca1d66e5bfd.aspx

Daily Mail (2010) Chinese UN Diplomat Apologises to Secretary-General for Tribute That Turned into Drunken Rant, *Daily Mail* (London), September 10. Available online at: www.dailymail.co.uk/news/article-1310810/Chinese-UN-diplomat-Sha-Zukang-apologises-Ban-Ki-moon-drunken-rant.html

Deiss, J. (2011) Informal Thematic Debate of the 65th Session of the United Nations General Assembly on Green Economy: A Pathway to Sustainable Development, New York, UN. Available online at: www.un.org/en/ga/president/65/initiatives/Green%20Economy/PGA%20summary%20on%20green%20economy%20thematic%20debate%20(2June2011).pdf

Dodds, F. (2011) High Level Dialogue on Strengthening the Institutional Framework for Sustainable Development. Available online at: www.uncsd2012.org/content/documents/SOLO%20DISCUSSION%20PAPER_TEXT.pdf

Dodds, F. and Nayar, A. (2012) Rio+20: A New Beginning, Nairobi, UNEP. Available online at: www.unep.org/civil-society/Portals/24105/documents/perspectives/ENVIRONMENT_PAPERS_DISCUSSION_8.pdf

Germany (2001) Water – Key to Sustainable Development: The Bonn Keys. Available online at: www.earthsummit2002.org/ic/freshwater/Bonn%20Keys.pdf

Glaser, G. and Bates, P. (2011) Enhancing Science–Policy Links for Global Sustainability, London, Stakeholder Forum. Available online at: www.stakeholderforum.org/fileadmin/files/ICSU%20Science%20Policy%20Final.pdf

Global Sustainability Panel (2011) Meeting Report, High-level Panel on Global Sustainability, Third Meeting of the Panel, Helsinki, UN, May 16–17, 2011. Available online at: www.un.org/wcm/webdav/site/climatechange/shared/gsp/docs/20110621%20Meeting%20Report%20-%20GSP3%20Helsinki%20public.pdf

Göpel, M. (2012) Ombudspersons for Future Generations as Sustainability Implementation Units, London, Stakeholder Forum. Available online at: www.stakeholderforum.org/fileadmin/files/SDG%204%20Ombudspersons%20for%20Future%20Generations%20Thinkpiece.pdf

Greenfield, O. (2011) Managing Natural Capital, London, Global Transition 2012. Available online at: www.stakeholderforum.org/fileadmin/files/PAPER%208_Managing%20Natural%20Capital_final.pdf

ICSU (2012) Future Earth, Paris, ICSU. Available online at: www.icsu.org/future-earth

IISD Reporting Services (2012) World Bank–IMF Spring Meetings Hold Informal Breakfast on Rio+20, New York, IISD, April 23. Available online at: http://uncsd.iisd.org/news/world-bank-imf-spring-meetings-hold-informal-breakfast-on-rio20/

Indonesia (2011) Solo Message, Solo, Indonesia. Available online at: www.uncsd2012.org/content/documents/Chairs%20Summary%20from%20Solo%20meeting.pdf

International Monetary Fund (2012) Development Committee communiqué, April 21. Washington, DC, IMF. Available online at: www.imf.org/external/np/cm/2012/042112a.htm

Ivanova, M. (2011) Global Governance in the 21st Century: Rethinking the Environmental Pillar, London, Stakeholder Forum. Available online at: www.stakeholderforum.org/fileadmin/files/IEG%20Paper-Ivanova-Final%20_2_.pdf

Jenkins, T. and Simms, A. (2011) The Green Economy, London, Global Transition 2012. Available online at: www.stakeholderforum.org/fileadmin/files/PAPER%201_Green%20Economy_Final_vj.pdf

Johnson, V. and Hälström, N. (2011) Energy Resources and Services: Achieving Universal Access to Energy and a Low Carbon, High Well-Being Economy, London, Global Transition 2012. Available online at: http://globaltransition2012.org/wp-content/uploads/2011/12/PAPER-6_ENERGY_Final_vj-1.jpg

Lynch, C. (2010) Departing U.N. Official Calls Ban's Leadership "Deplorable" in 50-Page Memo, *Washington Post*, July 20. Available online at: www.washingtonpost.com/wp-dyn/content/article/2010/07/19/AR2010071904734.html

Japan (2012) Tentative Chair's Note, Post-MDGs Contact Group, New York. Available online at www.mofa.go.jp/mofaj/gaiko/unsokai/pdfs/67_mdgs_se_gh_5.pdf

Khor, M. (2011) Perspectives on an Institutional Framework for Sustainable Development (IFSD). Available online at: www.uncsd2012.org/content/documents/khorsolo.pdf

King, M. (2009) King 3 Code of Corporate Governance for South Africa, Johannesburg, Institute of Directors. Available online at: www.iodsa.co.za/?page=kingIII

Horváth, K. L. (2012) A Preliminary Examination of the Possibilities for the Institutional Representation of the Interest of Future Generations and Environmental Protection within the System of the United Nations, London, Stakeholder Forum. Available online at: www.stakeholderforum.org/fileadmin/files/SDG%20Luca%20final.pdf

Kulish, N. (2011) Two Men Arrested in German Bōmb Plot, *New York Times*, September 8. Available online at: www.nytimes.com/2011/09/09/world/europe/09germany.html?_r=0

Lindner, C. (1993) The Earth Summit's Agenda for Change, Geneva, Centre for Our Common Future (reprinted 2005 in J. R. Strachan, G. Ayre, J. McHarry and R. Callway, *The Plain Language Guide to the World Summit on Sustainable Development*, London, Earthscan, p. viii).

Monaco (2011) The Monaco Message. Available online at: www.uncsd2012.org/content/documents/338Monaco%20Message.pdf

Nærstad, A. (2011) Food Security: The Earth Summit 2012 Should Give Strong Support to Small Scale Sustainable Agriculture, London, Global Transition 2012. Available online at: www.stakeholderforum.org/fileadmin/files/PAPER%207_Food%20Security_Final%20(2).pdf

Niestroy, I. (2012) Sustainable Development Councils at National and Sub-National Levels Stimulating Informed Debate: Stocktaking, London, Stakeholder Forum. Available online at: www.stakeholderforum.org/fileadmin/files/NiestroySDG%20thinkpiece%20-%20FINAL2.pdf

Oxfam (2012) A Safe and Just Space for Humanity: Can We Live within the Doughnut?, Oxford, Oxfam International. Available online at\; www.oxfam.org/en/grow/policy/safe-and-just-space-humanity

Parker, J. (2011) The 9 billion-people question, *The Economist*, February 24. Available online at: www.economist.com/node/18200618

PRME (2012) Global Forum – PRME @ Rio+20. Available online at: www.unprme.org/global-forum/prme-rio-20.php

Riddlestone, S. (2011) Living within Limits: One Planet Living, London, Global Transition 2012. Available online at: http://stakeholderforum.org/fileadmin/files/PAPER%204_One%20Planet%20Living_Final.pdf

Rockström, J., Steffen, W., Noone, K. *et al.* (2009) A Safe Operating Space for Humanity, *Nature*, 461: 472–475.

Rosemberg, A. (2011) Green Jobs and Skills: Decent and Green Jobs with a Just Transition: A Step towards Sustainable Development, London, Global Transition 2012. Available online at: http://docsfiles.com/pdf_challenge_paper.html http://globaltransition2012.org/wp-content/uploads/2011/12/SF-NEFpaper_Green-Jobs-1.jpg

Seaford, C., Mahoney, S., Wackernagel, M., Larson, J. and Gallegos, R. R. (2011) Beyond GDP: Measuring Our Progress, London, Global Transition 2012. Available online at: www.stakeholderforum.org/fileadmin/files/PAPER%202_BEYOND_GDP_Final_vj%20(2).pdf

Sha Zukang (2012) Sha Speaking at First Informal Negotiations on Zero Draft Jan 25 2012. Available online at: www.uncsd2012.org/content/documents/721sha4.pdf

Shrubsole, G. (2011) Trade: Outsourced Emissions and Global Trade in the Green Economy, London, Global Transition 2012. Available online at http://global transition2012.org/wp-content/uploads/2011/12/PAPER-11_Trade-and-Outsourced-Emissions_Final-1.jpg

SIDS Energy for All (2012) The Barbados Declaration. Available online at: www.sids energyforall.org/conference-details/decisions-agreements-and-declarations/

Stakeholder Forum (2008) Donostia Declaration, London, Stakeholder Forum. Available online at: www.stakeholderforum.org/fileadmin/files/Earth_Summit_2012/Earth_Summit_2012_Donostia_Declaration__2_.pdf

Stakeholder Forum (2012) Rio+20: Analysis of Zero Draft Submissions, London, Stakeholder Forum. Available online at: www.stakeholderforum.org/fileadmin/files/Rio%20Zero%20Draft%20Submission%20Analysis%20-%20FINAL.pdf

Strandaneas, J.-G. (2011) A Council for Sustainable Development: A Possible Outcome of the Rio+20 Process, London, Stakeholder Forum. Available online at: www.stakeholderforum.org/fileadmin/files/JGS-SDC-final.pdf

UN DESA (2012a) Sustainable Energy. Available at: www.uncsd2012.org/index.php?page=view&type=12&menu=153&nr=371&theme=3

UN DESA (2012b) Rio+20 Issues Briefs. Available online at: www.uncsd2012.org/resources.html

UN DESA (2012c) UN goes PaperSmart at Rio+20 to minimize environmental impact. Available online at: www.uncsd2012.org/content/documents/781rio20%20sustainability%20in%20action_press%20release%20final.pdf

UNECLAC (2011) (4) Rio+20: Sustainable Development Goals: Proposal by the Government of Colombia for Consideration by the Participating Countries, Santiago, UNECLAC. Available online at: www.eclac.org/rio20/noticias/paginas/6/43906/2011-613-Rio+20-Note_by_the_secretariat_Colombia_note.pdf

UNEP (2012) *Global Environment Outlook 5*, Nairobi, UNEP. Available online at www.unep.org/geo/geo5.asp

UN-HABITAT (2012) UN-HABITAT's Work for Rio+20 on Sustainable Cities and Urbanization. Available online at: www.unhabitat.org/categories.asp?catid=671&q=Rio%2B20

United Nations (1992) Agenda 21, New York, United Nations. Available online at http://sustainabledevelopment.un.org/content/documents/Agenda21.pdf

United Nations (2001) Road Map towards the Implementation of the United Nations Millennium Declaration, New York, UN. Available online at: www.un.org/millenniumgoals/sgreport2001.pdf?OpenElement

United Nations (2008) UN General Assembly, President of the 62nd Session: Addressing Climate Change. Available online at: www.un.org/ga/president/62/ThematicDebates/themclimatechange.shtml

United Nations (2011a) Membership, High-level Panel of the UN Secretary-General on Global Sustainability. Available online at: www.un.org/wcm/content/site/climatechange/pages/gsp/group-members_1

United Nations (2011b) "Chairs' Text" from the Sixty-Fourth Annual UN Department of Public Information Non-Governmental Organizations Conference, Bonn, September 3–5. Available online at: www.un.org/wcm/content/site/ngoconference/conference

United Nations (2011c) Compilation Document, New York, UN. Available online at: www.uncsd2012.org/compdocument.html

United Nations (2012a) Discussions on Zero Draft and Statements by Member States, Major Groups, NGOs, UN Agencies System, New York, UN. Available online at: www.uncsd2012.org/index.php?page=view&nr=333&type=12&menu=110&template=435 and www.uncsd2012.org/index.php?page=view&type=13&nr=409&menu=110

United Nations (2012b) Zero Draft, New York, UN. Available online at: www.uncsd2012.org/content/documents/370The%20Future%20We%20Want%2010Jan%20clean%20_no%20brackets.pdf

United Nations (2012c) *Resilient People, Resilient Planet: A future worth choosing*. Available at: www.un.org/gsp/sites/default/files/attachments/GSP_Report_web_final.pdf

United Nations (2012d) Oceans Compact. Available online at: www.un.org/Depts/los/ocean_compact/oceans_compact.htm

United Nations (2012e) The Secretary General: Leading on Climate Change, New York, UN. Available online at: www.un.org/wcm/content/site/climatechange/pages/gateway/secretary-general

United Nations (2012f) The Challenge. Available online at: www.un.org/en/zerohunger/challenge.shtml

United Nations (2012g) The Future We Want, New York, UN. Available online at: https://rio20.un.org/sites/rio20.un.org/files/a-conf.216l-1_english.pdf.pdf

United Nations General Assembly (2011) Protection of Coral Reefs for Sustainable Livelihoods and Development, UNGA 65/150. Available online at: www.icriforum.org/sites/default/files/UNGA_65_150.pdf

United Nations General Assembly (2012) Protection of Coral Reefs for Sustainable Livelihoods and Development, New York, UNGA 66/194. Available online at: www.un.org/ga/search/view_doc.asp?symbol=%20A/RES/66/194

UN-HABITAT (2012) UN-HABITAT's Work for Rio+20 on Sustainable Cities and Urbanization. Available online at: www.unhabitat.org/categories.asp?catid=671&q=Rio%2B20

UN News Centre (2012a) Rio+20: Ban Announces More than 100 Commitments on Sustainable Energy. Available online at: www.un.org/apps/news/story.asp?NewsID=42297#.Ui3zKdKsgdo

UN News Centre (2012b) Secretary-General Launches New Initiative to Protect World's Oceans. Available online at: www.un.org/apps/news/story.asp?NewsID=42668&Cr=ocean&Cr1=#.UYXfmr_3C-s

UN-NGLS (2011) UN Secretary General Announces His Priorities for his Second Term. Available online at: www.un-ngls.org/spip.php?page=article_s&id_article=3643

Watts, J. (2012) Rio+20 Organisers Struggle to Untangle New World Disorder, *Guardian* (London), June 14. Available online at: www.guardian.co.uk/environment/2012/jun/14/rio-20-conference-new-world

Woodward, D. and Abdallah, S. (2011) Global Inequality, London, Global Transition 2012. Available online at: www.stakeholderforum.org/fileadmin/files/PAPER-3_global-inequality_Final_vj-1.pdf

World Economic Forum (2011) *Water Security: The Water–Food–Energy–Climate Nexus*, Geneva, World Economic Forum. Available online at: www3.weforum.org/docs/WEF_WI_WaterSecurity_WaterFoodEnergyClimateNexus_2011.pdf

World Future Council (2012) Adding Our Voice to SEE4ALL Civil Society Statement. Available online at: http://power-to-the-people.net/2012/06/sustainable-energy-for-all-wfc-signs-civil-society-statement/

Zadek, S. (2011) Shaping a Public Fiduciary: Aligning Financial Markets to Sustainability, London, Global Transition 2012. Available online at: www.stakeholderforum.org/fileadmin/files/PAPER9_simon_zadek.pdf

Chapter 3

# How did it all come to happen?

## The decisions that were required in order to start building a new development framework

This chapter will take the reader through the different stages of the negotiating process and through the decisions that were required, from all actors involved, in order to allow the world leaders gathered in Rio de Janeiro to adopt the official outcome of the conference – the political document entitled "The Future We Want" – and to launch a significant number of important initiatives on sustainable development. It will start by pointing out the key divergences that remained as the meeting was about to start, and also some of the hurdles that made progress difficult. Its second section will point to the demanding work invested by developing countries gathered under the umbrella of the Group of 77. It will also explain the efforts undertaken since the beginning of the Rio+20 preparations into building a common vision by developing countries towards the conference and towards the text of "The Future We Want," throughout the negotiating process under the leadership of Ambassadors John W. Ashe and Kim Sook. This chapter will also reflect on the strong leadership exercised by the Brazilian authorities into finalizing, through a process of transparent yet focused consultations, the remaining gaps in the document. The third and last section will highlight the fact that the success of Rio+20 – "The Future We Want" and its multiplying outcomes and the different streams it allowed to flourish – is also due to valuable efforts made by individuals from different backgrounds and perspectives, all of which represent the diversity of the United Nations.

## Key divergences as the meeting started

UN headquarters conference room no. 2 was full with country negotiators, UN staff, international organizations representatives and stakeholders who had gathered on that Saturday evening of June 2nd to listen to the conclusions of the leaders of the Rio+20 Preparatory Committee, John W. Ashe and Kim Sook, on the results of the last stretch of negotiations before the start of the conference.

The co-chairs admonished the delegates: the entire effort for which they, the member states, the Secretary-General and the United Nations at large had been exercising was on the brink of failing. Over the previous week, the advances in the exercise of drafting an acceptable document for the conference had been erratic and yielded only a fraction of the progress required. Up to that point, most negotiators had accepted the idea that some of the key decisions and compromises that were vital for allowing Rio+20 to have a tangible outcome would require a higher level of political – perhaps ministerial – engagement in the process. This was particularly true for core issues such as the future of the United Nations Environment Programme (UNEP), the replacement of the Commission on Sustainable Development (CSD) and the modalities for the Sustainable Development Goals (SDGs). The negotiating settings in New York and the format that had been agreed by UN member states to conduct the negotiations were simply not the right place to reach political compromises on those matters. Up to the beginning of May, negotiators worked tirelessly under the assumption that by putting in extra hours in the preparatory process they would be able to take to Rio a workable text with only a few open paragraphs pending political agreement for a last-minute deal. However, despite all efforts this was not happening. Between April 23 and May 4, negotiators had reached agreement on only 21 out of a total of 420 paragraphs in the draft outcome document. By June 2, 70 additional paragraphs had been painstakingly agreed, yet 259 remained "open."

The gravity of the situation in early June should not be underestimated. The nineteenth session of the CSD (May 2–13, 2011) had failed, partly because of a disagreement in only two outstanding paragraphs. With 259 open paragraphs, there were 259 possible reasons for failure of the final outcome document. The dynamics of the two main working groups and the dozen thematic "splinter groups" had failed to achieve their most important purpose: preparing a practically finished draft political document that would reflect the best efforts by negotiators. After all, throughout the whole process the main negotiating groups, consisting of the European Union, which includes 28 member states, and the Group of 77, which comprises 132 member states plus the State of Palestine, had emphatically reaffirmed that Rio+20 was a "member-state driven process" and claimed to retain "ownership" over the text of the outcome document. These assertions discouraged the presentation of a third draft compromise proposal under the authority of the co-chairs.

The working methods used towards the end of the Rio+20 negotiations had huge negative consequences. The use of large-screen projectors to display on-screen the text as it was being redrafted and the authors of those amendments was drawn from negotiation experience in small, technically oriented environments (Chasek and Wagner, 2012). The transposition to large plenary settings for the sake of transparency was not successful. Instead of empowering the leading figures of the process and the skilled and widely trusted facilitators of each of the contact groups, and ultimately all negotiators, into building early agreements, these working methodologies tended to drive all participants to focus their attention mostly on the details of the text, the choice of adjectives and, in some cases, on issues of pride rather than substance.

The negotiating sessions in New York started with all member states assembled together and negotiating over the entire text, paragraph by paragraph, punctuation mark by punctuation mark. Once this approach had been seen to be tedious and time-consuming, in an effort to accelerate the process two negotiating groups, each led by one of the co-chairs, were set up: Working Groups 1 and 2. John W. Ashe led on Working Group 1, which focused on sections V, "Framework for Action," and VI, "Means of Implementation." Kim Sook led Working Group 2, which focused on sections I, "Common Vision"; II, "Renewing Political Commitment"; III, "The Green Economy"; and IV, "Institutional Framework for Sustainable Development."

With little progress being made, a number of so-called splinter groups were established by the co-chairs in order to discuss and move forward some of the more complex and controversial issues. The splinter groups were sub-groups, each with responsibility for trying to secure agreement on acceptable language for various parts of the text or a problematic thematic issue. The co-chairs decided on the splinter groups' discussion topic and leader, on the basis of member states' interests and the experience of their lead negotiators. Unfortunately, the word "splinter" also characterized the negotiations, their process and progress. The traditional problems of the slow pace of negotiating groups, lack of agreement on critical issues and bickering over brackets, paragraphs and language haunted the process until the last day of the New York-based negotiations and would continue to do so in Brazil right up until the day when the host country took over. Repeated attempts by Ban Ki-moon, by the conference secretary-general Sha Zukang and by the co-chairs, and the addition of an extra week of negotiations in New York, had all failed to get delegates to quicken their pace.

The Zero Draft had grown from 21 pages in its original submission to 6,000 pages after input from stakeholders. Ambassadors Ashe and Kim developed a Co-Chairs' Text of some 80 pages out of this mammoth document, on the basis of discussions with member states, the splinter groups and the early negotiations, which revealed the priorities and sticking points for member states. This new text was intended for use during the final weeks of negotiations and, although the co-chairs believed it would be acceptable to member states, the slow negotiation and review of every paragraph continued unabated. During the last week of negotiations on the Co-Chairs' Text in late May, the number of paragraphs in dispute swelled and shrank, but mainly swelled. Left to the hands of delegates, the co-chairs' 80-page text had swollen to 206 pages, with 401 paragraphs bracketed and a further 21 paragraphs agreed. On the closing day of negotiations in New York, 70 paragraphs had been agreed *ad referendum*, with a further 259 remaining to be agreed.

There is another element that became evident during the negotiations: the deep divide between developed and developing countries. Positions taken by the

European Union, the United States, Canada and Japan on the one hand, and by the Group of 77 on the other, were not easily reconcilable. Another problem that was appearing with greater frequency in multilateral negotiations also arose in the Rio+20 negotiations: growing internal tension within the Group of 77 resulting from the divergent interests among increasingly diverse countries. When the New York negotiations ended, countries in the Group of 77 were in a state of flux. The International Institute for Sustainable Development (IISD) made this telling commentary in its *Earth Negotiations Bulletin* – a trusted source for all sustainable development experts – in the final week of negotiations in New York:

> The unintentionally but aptly named "splinter groups" proliferated to over 20, precluding participation of small delegations, giving rise to complaints and consequently a feeling of denial. Many felt that the groups were convened too late in the process. The inability of many delegates to attend all of the groups, the dearth of strong facilitators and Secretariat support resulted in confusion on what had actually been debated and agreed. Indeed, in several instances during the closing hours on Saturday, delegations asked to re-bracket ostensibly agreed passages, just to be on the safe side. One long-time participant commented that an unexpected factor – information technology – had hit the preparatory process with a vengeance: instant communication led to "excessive" 24-hour control from capitals, sapping negotiators' initiative and slowing down negotiations. "Rarely has a drafting process been so erratic," observed one delegate. The disorder anesthetized the sense of urgency with some delegations acting as if they had months to go. Yet time was slipping away like sand in an hourglass.
>
> (IISD, 2012a)

Admittedly, there were substantive reasons for the disarray and snail-like pace of drafting. The first was the enormous range and sheer complexity of issues addressed in the draft outcome document, embracing as it does the three pillars of sustainable development. The second was the lack of early consensus on the parameters of the main issues on Rio+20 agenda: the green economy, the institutional framework for sustainable development (IFSD) and the more recently introduced Sustainable Development Goals (SDGs). As expected, these issues plagued negotiations in New York despite the new Co-Chairs' Text, impeding an improvement on previous drafts based on realistic assumptions of what might be acceptable to all.

At the conclusion of the meeting, the media reported that, with less than three weeks to go before the arrival of more than 100 world leaders in Rio de Janeiro, key areas of divergence remained. This was the case for several issues within the "Framework for Action," such as climate change, oceans, food and agriculture, migration, sexual and reproductive rights, biodiversity, and eco-system services, many of which – with the notable exception of the section on oceans – were largely based on previously agreed commitments and references.

Divergences also persisted in the process for the establishment of the SDGs, on the principles on which these should based and in the discussion as to whether there should be indication of initial "thematic areas" for the SDGs. Agreement was still lacking also on all of the key areas of the "Means of Implementation" section, most notably finance for sustainable development and on the crafting of a technology transfer interface, not to mention disagreements in the two centrepieces of the conference: the IFSD and the green economy. The reference to the use of indicators was also problematic. Up to a certain point there were 18 mentions of the use of data recollection to track progress; such a pronounced emphasis, however grounded in more than two decades of implementation of sustainable development policies, nevertheless generated suspicion that a new form of conditionality in the international cooperation framework or even in the definition of the paradigm of sustainable development could emerge from Rio. After all, three weeks before the conference started, the most telling points of divergence and uncertainty with regard to the future of the paradigm of sustainable development remained: were UN member states going to reaffirm their commitment to all the principles of the 1992 Rio Declaration on Environment and Development, including the principle of "common but differentiated responsibilities"? And were they going to reaffirm their desire to "fully implement" all of those commitments previously agreed to in the major UN conferences of the 1990s and the 2000s on the social, economic and environmental fields?

## The group of developing countries: too diverse to negotiate as a single block?

The Group of 77, created in June 1964, is one of the most significant political phenomena of the Cold War era. More than two decades after the end of the Cold War, it has remained as the most enduring coalition for multilateral international negotiations, giving to its members – which through the years have increased in number from 77 to 132 countries of the global South – a common voice and a common platform in all major UN conferences in the social, economic and environmental fields (Ahmia, 2006). The Group has been the single most relevant actor behind the UN summits on sustainable development since the 1992 Earth Summit. Along the years, however, the Group of 77 has managed to adjust and evolve its mechanisms for defining its own positions. Such evolution has occurred in response to increasing social, economic and political diversity among its members. The Group of 77's strengths in sustainable development negotiations have relied on its cohesion, on the principle of mutual solidarity and support, and, finally, on the remarkable individual capacity of each of its leading negotiators who have to date undertaken the burden of achieving results on behalf of their group of countries.

All of these factors have provided powerful incentives for retaining the Group's unity, spread across all ends of the development spectrum, ranging from major emerging industrialized economies to highly vulnerable Small Island Developing States. The main constituencies within the Group of 77 are: (1) the Small Island Developing States (SIDS); (2) the least developed countries; (3) the landlocked developing countries; (4) the African countries; (5) ASEAN; and (6) the Community of Latin American and Caribbean States (CELAC), which unites the developing countries of the Americas, with the exception of Mexico. Within these constituencies are sub-groups or coalitions of countries, all telling of the diversity of interests among countries of the developing world.

Throughout the Rio+20 negotiations, and owing to the able leadership of Argentina and Algeria, the Group of 77 maintained its cohesion, while allowing members to express the diversity and heterogeneity of their national positions. This was not an easy task. Rio+20 ran parallel to the Copenhagen (2009), Cancún (2010) and Johannesburg (2011) climate change conferences, during which the cohesion of the Group was often challenged. The Group of 77 remained together and managed to achieve a common position on practically all the issues of the Rio+20 Conference, despite the emergence and visibility acquired by the multiple cross-regional and thematic initiatives and groups active during the informal preparatory process of Rio+20 (covered in Chapter 2). There were wide divergences in some of the national and regional positions, which were openly expressed in the statements made in the plenary sessions by the constituencies represented within the Group.

Overall, the dynamics within the Group of 77 during the Rio+20 preparatory processes can be divided, for the reader's sake, into four different phases. The first phase was guided by a growing sense of collective ownership towards the conference. The second was illustrated by the transition from a "negotiating modality" to the definition of a common position. During the third phase the Group and its negotiators led the process and engaged with all other member states on the basis of a strong negotiating mandate. During the final stretch, the Group of 77 collectively closed ranks behind Brazil, once the host country took over the process.

## *The definition of a vision and the call for a framework*

In September 2008 the Group of 77 embraced the Brazilian initiative to host in 2012 a "World Summit on Sustainable Development." Its members' initial view was that the summit gave an opportunity to "review the 20-year progress achieved in the implementation of the outcomes of the United Nations Conference on Environment and Development" (G77, 2008). From that moment on, the Group of 77 occupied the driving seat for drumming support for the conference, and with its successive leaders – Antigua and Barbuda (2008), Sudan (2009), Yemen (2010) and Argentina (2011) – it significantly contributed to shaping the growing sense of developing country ownership towards the conference.

The Argentinian presidency had the additional responsibility of steering the Group into a "negotiating modality." In March 2011 the Group of 77 set out to define its own position. After nine months of continuous internal discussions, the Group's submission to the Zero Draft of "The Future We Want" was formally presented on November 1 of that year. Its initial vision on what could be achieved in Rio was divided into four broad sections: the objective of the conference; the green economy in the context of sustainable development and poverty eradication; an institutional framework for sustainable development; and an annexe on "conservation and sustainable use of fish stocks and other issues related to marine biodiversity and oceans."

The level of detail achieved by the Group of 77 in the first sections of its initial submission was not matched by a similar focus on the two themes of the conference: the green economy and the institutional framework. On the green economy, the Group of 77, although cognizant of the lack of a consensual agreement on its definition, managed to advance a few initial tenets. The green economy, from its perspective, involved undertaking economic and social activities in an environmentally sound manner; ensuring social inclusion and equity; including tools to catalyse international cooperation for achieving sustainable development in areas such as technology transfer, sustainable consumption and production patterns; and (in a cautious tone) "analyzing the effects of practices" for quantifying and evaluating ecosystem services as well as for establishing market mechanisms for such services. On the institutional framework for sustainable development, the Group of 77 highlighted two overall functions that institutions should perform: the implementation of sustainable development and the integration of its pillars (social, environmental, economic). The Group also agreed 15 system-wide institutional functions that complemented these broad objectives.

The initial difficulty of defining, at the beginning of the process, concrete proposals for both themes within the Group was not the result of a negotiation strategy. The negative economic trends and the emerging actors highlighted in Chapter 1 illustrate why the Rio+20 preparatory process was a particularly challenging exercise for developing countries. If we look back to the documents agreed at each of the regional consultations for Africa, Latin America, Asia and the Arab states, it is clear that developing countries had a detailed perception of how to advance (or not) the green economy and the institutional framework. The challenge was to unite these different visions under a common position.

## *Consensus-building among developing countries*

The question remaining for the Algerian presidency of the Group of 77, as it took over at the beginning of 2012, was, therefore, how to reconcile divergent visions under one common approach. The decision by Algeria to lead the group of developing countries, particularly in midst of the evolution of the Arab Spring, was an important step for its foreign policy. The Algerian commitment to the Group of 77 was swiftly endorsed by the African Union; this allowed the incoming presidency to work jointly with Argentina on some of the initial

*Figure 3.1* The Group of 77 consulting on the green economy during the second round of informal consultations. Pictured are Josefina Bunge from Argentina (centre), Rene Orellana from Bolivia, and Mario Rubiales from Ecuador (courtesy of IISD/*Earth Negotiations Bulletin* (www.iisd.ca)).

political decisions that were taken for the organization of the preparations for the Rio+20 Conference. By early January 2012, when the Zero Draft of "The Future We Want" was distributed, the Group started the demanding exercise of striking difficult compromises.

One of the key initial issues where the Group of 77 spent time in reaching consensus was how to reflect and highlight in a balanced way the sustainable development challenges faced by the different constituencies represented within the Group. It was resolved through reaching consensus on separate sections dedicated to Small Island Developing States, least developed countries, land-locked developing countries, African countries and middle-income countries.

As negotiations progressed on the basis of the Zero Draft of "The Future We Want," which included a thematic framework for action, the Group of 77 was confronted with the need to define, in the light of different national perspectives, common positions on a host of challenging issues. Some of the most complex issues for the Group were the right to water and water cooperation; oceans and seas; sustainable energy; climate change; desertification and land degradation; biodiversity and ecosystem services; sexual and reproductive rights; as well as the references to governance and human rights. Sustainable consumption and production (SCP) and the adoption of a ten-year framework of programmes on

SCP were one of the main areas where the group found a common ground. All of the diverging issues were eventually solved by working through small groups composed essentially of interested delegations.

With regard to the institutional framework, the Group of 77 worked until the last weeks of the preparatory process to define a common position on the replacement of the CSD and on the strengthening and upgrading of UNEP. There were strong divergences even among those constituencies where positions seemed in principle more defined. This was the case for the African countries. The 2011 consensus statement of the African Union did not allow a definitive interpretation on the extent of the transformation of UNEP. The African position described the transformed UNEP as an "international specialized institution," thus falling short, from the perspective of some countries, of interpreting this as a call for a treaty-based specialized agency, as defined in article 57 of the UN Charter. Divisions were not confined to the African region: several Latin American and Arab countries were openly sceptical about the proposal of transforming UNEP into a fully fledged specialized agency despite the fact that the proposal had gained many supporters from all regions.

The proposal to establish a Sustainable Development Council was also divisive within the Group of 77. It is possible that its universal character attracted some countries within each of the sub-groups of developing countries. They thought that a new institution could overcome the lack of implementation of the decisions taken by the CSD. It was also thought that previous reform efforts by the Economic and Social Council (ECOSOC), specifically after the 2005 World Summit Outcome, had not been effective enough to suggest that this organ could take on board the functions of the CSD. Finally, sceptical countries within the Group pointed to the fact that ECOSOC had limited reform potential, since it was bound by the provisions of the UN Charter, which gave it a limited membership of 54 states. Other countries within Latin America, Africa and Asia strongly favoured giving ECOSOC a central role in the integration of the three dimensions of sustainable development. They thought that it had already important functions in the social pillar, particularly through its Commission on Social Development in the follow-up to the Monterrey Consensus, and the dialogue with the Bretton Woods institutions. Above all, the Council was in charge of the following up of the MDGs and of the coordination of the sprawling UN operational activities for development. Such arguments made a compelling case for giving ECOSOC a "key role" in the integration of the three dimensions of sustainable development. Cuba took the lead in this regard by putting forward at the beginning of the preparatory process a detailed proposal to establish a Global Sustainable Development Forum which would combine, in a new body subsidiary to ECOSOC, the strengths and added value of, first, the Global Ministerial Environment Forum convened by UNEP and, second, of the CSD. It introduced the additional innovation of proposing a joint secretariat between UN DESA and UNEP to service the Forum.

On the green economy and sustainable energy themes, the Group of 77 was also confronted with different, albeit more nuanced, views. Some countries, particularly some energy-producing ones, along with other middle-income countries, were openly concerned about the burden imposed by a "transition towards the green economy." Other countries, within Asia, the Caribbean and Africa, not only suffered from the high costs of importing energy but also had embraced green economy approaches within their own national development strategies. For these reasons the group had to make difficult compromises, which in the end were beneficial to both the green economy and the sustainable energy sections of "The Future We Want." On the green economy in particular, the internal discussions within the Group of 77 were critical in giving the notion of the green economy – hitherto characterized by essentially economic notions – a broader and richer application which emphasizes the social and inclusive character that all green economy policy approaches should have.

### *Moving towards a unified position*

From the first weeks of the preparatory process until the last minutes of the Conference, the priority within the Group of 77 was to identify sufficient common ground to keep the Group united. The strategy followed on the most contentious issues which influenced the final shape of "The Future We Want" was to build a bottom-up understanding on the basis of the agreement on the functions and principles for each of the four core outcomes of the conference: the green economy, the strengthening and upgrading of UNEP, the establishment of a High-Level Political Forum and the definition of Sustainable Development Goals. This strategy, devised by the Algerian presidency, proved in perspective successful. By early May, after spending an estimate of 500 hours of internal discussion, the Group of 77 had managed to reach a common position on almost all key issues.

On the green economy, the Group reached a consensus on three key principles that from then on were to guide its approach in the negotiations. The first principle was that there were different visions and approaches towards the green economy. The second was that the green economy was to be defined only as a tool through which to achieve sustainable development. The third was to maintain an overall constructive approach towards the green economy so as to allow the necessary flexibility for those developing countries wishing to adopt green economy strategies. The Group of 77 was never in the position of accepting the notion of a "universal transition" towards the green economy or of agreeing a timeline for such a transition.

The second core issue agreed within the Group of 77 was the strengthening of the environmental pillar. Leaving aside the name and specific institutional identity, the Group worked on reaching agreements on the basis of the necessary functions of enhanced international environmental governance. Through this exercise it became gradually clear to most of the negotiators that their countries were united in their common desire to have a more efficient and stronger environmental pillar, endowed with increased financial resources and capacities to respond to national implementation needs. Developing countries also favoured achieving "greater policy coherence" among multilateral environmental agreements (MEAs). These basic aspirations were expressed in the G77 "final compromise text" dated May 2, a month ahead of the conference. It proposed to establish universal membership within UNEP's Governing Council in order to "allow equal participation of all states in its deliberations and give it authority to lead and set the global policy agenda for the environment." It also contemplated improving "UNEP's efficiency and operational capacity through the provision of more secure, stable, additional and predictable financial resources from the regular budget and voluntary contributions." The Group's position reaffirmed UNEP's coordination mandate within the United Nations and the importance of strengthening and consolidating its presence in Nairobi. Finally, the Group of 77 position acknowledged "the need to address fragmentation of the secretariats of environmental conventions, platforms, panels, initiatives and mechanisms including the financial mechanism within the UN," thus opening a door through which to tackle the central question of international environmental governance: how to build coherence and complementariness among legally autonomous environmental regimes.

On the sustainable development arrangements, the Group of 77 followed a path similar to the one used for environmental governance and the green economy. The Group recognized that ECOSOC had "a role for achieving a balanced integration and implementation of the three dimensions of sustainable development," thus radically departing from the traditional view still dominant a few years before. It also proposed that Rio+20 could decide, in such context, to establish "a high-level political forum with intergovernmental character, building on existing relevant structures or bodies including the CSD" (IISD, 2012b). The concept of the high-level political forum drew from the "non-meetings" convened by the Mexican ambassador, Luis Alfonso de Alba, covered in Chapter 2, as well as from the Cuban proposal for a global sustainable development forum.

The compromise text acknowledged that there was no agreement within the Group of 77 on the relationship such a high-level forum would have with the UNGA and ECOSOC; it did specify that its purpose and mandate were to follow up on the sustainable development agenda as defined in 1992 in Agenda 21 and further elaborated in the Johannesburg Plan of Implementation, the Barbados Plan of Action and the Mauritius Strategy. The position of the Group specified that the new forum was meant to be universal, while allowing system-wide participation and conducting its work under more flexible and "action-oriented" working methods than those used by the CSD. It was also expected to enhance accountability, to evaluate commitments related to means of implementation and to facilitate the provision of financial resources, without leading, however, to the establishment of a periodic review mechanism.

However, such a review mechanism was actively sought as a key deliverable by European, particularly Nordic, countries.

Despite having been initiated by members of the Group, the proposal to have Sustainable Development Goals was introduced within the Group of 77 after the release of the Zero Draft of "The Future We Want." It took several consultations before the Group embraced this transformative idea. This endorsement was achieved for the most basic elements of the SDGs (have the goals, set out basic principles to define the goals and define a mandate for the UNGA to decide upon these) by the third intercessional meeting (March 26–27, 2012). Although a few countries advocated that concrete areas or even specific goals be advanced, Algeria, in its capacity of chair of the Group of 77, supported by many other countries, remained reluctant to accept this idea. With only one month to go before the start of the conference, the chair felt that it was simply unrealistic to undertake at that stage a serious attempt to define targets, goals or even thematic areas without opening a Pandora's box.

This reluctance was not grounded in time constraints alone. Developing countries were the first ones to realize in the last stretch of the Rio+20 preparatory process that a fragmented development agenda (between MDGs and SDGs) would have resulted in the dispersion of their own national development efforts. A key question for the vast majority of developing countries was how to agree to SDGs during a global economic recession without compromising the achievement of the MDGs. It was not only a matter of availability of financial resources for international cooperation. After more than a decade of implementation of the MDGs, most developing countries agreed that this framework had helped them to promote coherence within their own national poverty eradication strategies. For the Group of 77, the MDG framework had "shaped a broad development vision" and had also ensured international policy coherence. The world could not move its attention away from fighting poverty. As indicated by March 28, 2012 in the Co-Chairs' Text, paragraph 105:

> We recognize that, three years from the 2015 target date of the MDGs, while there has been progress in reducing poverty in some regions, this progress has been uneven and the number of people living in poverty in some countries continues to increase, with women and children constituting the majority of the most affected groups, especially in least developed countries and particularly in Africa.
>
> (United Nations, 2012)

The position of the Group of 77 on the SDGs opened a door for an integrated sustainable development agenda. The SDGs, equally built upon the MDGs and the principles contained in the Rio Declaration, could be a driver for implementation and universal mainstreaming of sustainable development, as well as of integration of its three dimensions. The common position of the Group of 77 allowed its negotiators to lead the process and engage with their partners on the basis of a strong political mandate. This was particularly evident from the outset of the second round of informal negotiations starting late April 2012 until the last hours of the third preparatory committee meeting, which closed on June 15. During this period, a few extremely capable negotiators drawn from the entire Group of 77 constituencies led in the negotiations.

### *A new generation of developing country leaders*

Rio+20 marks a complete generational overhaul from the developing country representatives who had led the Group of 77 in Rio 1992 or in Johannesburg 2002. A new generation of representatives from the South undertook the task of crafting a common position and finding sensible compromises for all of the key issues of the conference. This was a substantive difference from other countries and negotiating groups such as the European Union, whose positions were in many cases informed by senior officers who had been involved in the Johannesburg or the Rio conferences. These negotiators from the South are characterized by being career diplomats with a high degree of academic specialization, in some cases in countries other than their home countries. A last distinguishing element, besides their common attachment to the principles of the United Nations, is their close involvement with the climate change negotiations under the UNFCCC and their continuous engagement in the Post-2015 Development Agenda. This sub-section attempts to bring to light the contributions of a selection of these representatives from developing countries to Rio+20 and, in doing so, aims to recognize through their efforts the willingness of all of the Rio+20 negotiators to deliver a coherent and integrative vision for sustainable development, one that ultimately captures the complexities among the developing world.

Abdelghani Merabet was the Algerian chief negotiator who chaired, during his country's presidency of the Group of 77, all the expert preparatory meetings that defined the position of the Group of 77 ahead of Rio+20. He and his ambassador, Mourad Benmehidi, are credited with maintaining the group's integrity throughout the long negotiation process for Rio+20 and for wisely managing the diverse and often competing interests within the countries of the Group. Abdelghani Merabet is a career diplomat who also served as vice-chair of the Commission on Sustainable Development during its nineteenth session (2011).

Rueanna Haynes is a career diplomat from the Republic of Trinidad and Tobago. She is a lawyer by training and a graduate of the University of the West Indies and the Hugh Wooding Law School in Trinidad and Tobago. During the preparatory process for Rio+20, she successfully facilitated the negotiations on the SIDS and regions sections of "The Future We Want." Her main contribution in facilitating this discussion was to ensure that a balanced treatment was given to countries in special situations, including SIDS, LDCs and African states, while ensuring that the sustainable developmental challenges of specific regions and

*Figure 3.2* Natalia Novoa from Uruguay, Rebecca Weber from the United States, Alexandra Dedu from the European Union and Ulf Melgaard from Denmark try to find a last-minute compromise on the institutional framework during the informal consultations led by the host country (courtesy of IISD/*Earth Negotiations Bulletin* (www.iisd.ca)).

regional efforts could also be reflected. Through Rueanna Haynes' facilitation, the final text was a compromise that struck the appropriate balance of interests for all parties present. In addition to this accomplishment, Haynes has been a leading negotiator on behalf of the Alliance of Small Island States (AOSIS) in the climate change processes under the UNFCCC and the Kyoto Protocol.

Mohamed Khalil joined the Egyptian Ministry of Foreign Affairs as a diplomat in 1997 and so had 15 years of experience in the field of diplomacy across different areas, including international politics, economic affairs and sustainable development. He served in two diplomatic posts in Africa and Europe before joining the Egyptian mission to the United Nations in 2010. Mohamed Khalil negotiated many sections of "The Future We Want" on behalf of the Group of 77. However, his contribution was not limited to the preparations for Rio+20; after the conference he followed up, on behalf of his country and the Group, the commitments made in Rio. He led the negotiations on Resolution 67/203 on the implementation of Rio+20, which made it possible to kick-start the process for the implementation of the ten-year framework of programmes on sustainable consumption and production patterns and the workshops under the UNGA on a technology facilitation mechanism. He also led, during 2013, the negotiations for the establishment of the High-level Political Forum. Like his colleagues, Khalil has substantive experience in the climate change negotiations.

Natalia Novoa joined the Ministry of Foreign Affairs of Uruguay in 2005 and was posted in 2008 to her country's mission to the United Nations. From that position she developed her expertise in most of the developmental and international cooperation issues covered by the UNGA and the Economic and Social Council, such as trade, middle-income countries, financing for development and operational activities of the United Nation. Natalia Novoa's main contribution to Rio+20 was the drafting during 2011, together with Ana Bianchi from Argentina, of the Group's initial submission for the institutional framework section of the Zero Draft of "The Future We Want." From that basis, Natalia led the group's negotiations on that critical section of the outcome document.

Farrukh Khan joined the permanent mission of Pakistan to the United Nations in New York in October 2010 following his earlier assignments in his own country and in Geneva. He joined Pakistan's diplomatic service in 1994. During the crucial last weeks of Rio+20, Farrukh Khan negotiated on behalf of the developing countries the final agreement that made it possible to chart a path forward for the SDGs. Shortly after Rio, in February 2013, he made a second fundamental contribution by presiding, together with Mexico, over the negotiations at the first universal session of the Governing Council of UNEP, which led to the establishment of the United Nations Environment Assembly of UNEP, a fundamental outcome of Rio+20. With more than 15 years of diplomatic experience, Farrukh Khan specializes in three interlinked issues: sustainable development, energy and climate change. He is an elected alternate member of the Green Climate Fund Board. He was also elected as a member of the Transitional Committee that designed the Green Climate Fund (GCF), where he led the process of establishing a private-sector facility inside the GCF's board.

Jimena Leiva Roesch is the sustainable development expert from the Guatemala mission to the United Nations. She is passionate about advancing the global sustainable development agenda. During the Nineteenth Session of the CSD in 2011, Jimena Leiva Roesch, together with a leading diplomat from Singapore, Wu Ye-Min, led the group of developing countries in reaching an ambitious compromise for the ten-year framework of programmes on sustainable consumption and production patterns, one that would allow its final adoption at Rio+20. Leiva Roesch, like most of her colleagues highlighted in this section, has represented the voice of the developing world in the climate change negotiations.

Josefina Bunge is an Argentinian Foreign Service officer who has witnessed many negotiations within the UN system. She was appointed as a vice-chair of the Bureau for the Preparatory Process of the United Nations Conference on Sustainable Development, representing the Latin American and Caribbean region. During Rio+20 she represented and promoted consensus in the Group of 77 on the discussions around the green economy.

Thembela Ngculu, a South African diplomat, was posted in 2010 to his country's mission to the United Nations in New York as a counsellor for sustainable development. He was instrumental during the Rio+20 negotiations in helping to raise the level of ambition within the green economy section of "The Future We Want," as well as on the strengthening of the environmental pillar of sustainable development.

Syahda Guruh L. Samudera was posted in 2011 to Indonesia's permanent mission to the United Nations. He had the excruciating task of representing the Group of 77 during the negotiations of most of the thematic areas and cross-sectoral issues contained in section V of "The Future We Want." After Rio, he joined Indonesia's team supporting the country's President, Susilo Bambang Yudhoyono, as co-chair of the High-Level Panel of Eminent Persons for the Post-2015 Development Agenda. Like his colleagues, Syahda Guruh L. Samudera has been part of his country's delegation at the climate change negotiations in Durban 2011 and Doha 2012, where he specialized on issues related to the legal aspect of the post-2015 climate change regime.

## The European Union and Rio+20

The European Union (EU) had been reluctant to have a Rio+20 Conference since the proposal had been initially floated. It was initially unclear about what the added value would be and feared it might become another EU/donor-bashing event. This opposition eventually broke when the President of France, Nicolas Sarkozy, visited Brazil on September 6–7, 2009 and announced unilaterally his support for a Rio+20 Conference:

> France fully supports the desire of Brazil to organize in 2012 a new Earth Summit; 20 years after the first Summit had been held in Rio de Janeiro. This new summit may be an ideal opportunity to make decisive and necessary progress for the international governance of the environment.
>
> (Sarkozy, 2009)

The European Union's original hopes for Rio+20 were for a set of concrete commitments on sustainable development around the green economy and the upgrading of UNEP into a World Environmental Organization (WEO). Germany and France had advocated this latter position for over ten years. However, not all EU countries agreed with it, and they could be heard playing down the importance of the idea in bilateral meetings. Particularly lukewarm was the United Kingdom, which under consecutive governments had not been convinced of the need for the upgrade, or the possibility of achieving it.

In its submission to the Zero Draft in November 2011, the European Union identified a set of "ingredients" for the overall outcome:

- a broad political "rallying call" with a shared, ambitious vision and goals;
- a set of specific actions at international, regional and national level, mapped out as a "green economy road map";
- a "toolbox" of policy approaches and best-practice examples to be used to reach agreed objectives;
- a mechanism for promoting and monitoring overall progress.

The Union said:

> A Green Economy Roadmap can guarantee continued commitment beyond Rio+20, ensuring that the agreed vision and goals will be followed through in a systematic manner. It can map out a range of international, regional and national actions with milestones, indicators and targets, as well as mechanisms to monitor overall progress.
>
> (European Union, 2011)

The Union was promoting a transition to a green economy that included investing in the sustainable management of key resources and natural capital, and establishing the right market and regulatory conditions, meaning that environmentally harmful subsidies should be addressed. This became a very controversial approach, as developing countries viewed it as "imposing" a top-down approach to the green economy. ALBA (Alianza Bolivoriana para los Pueblos de Nuestra América, or Bolivarian Alliance for the Peoples of Our America; see Chapter 1 for further information) country delegates in particular saw it as selling nature through green grabbing. This attack on the basis of the "commodification of nature" had many supporters within the social movement, indigenous peoples and some NGOs as well.

By March 9, 2012 the Council of the European Union had reaffirmed its call for a green economy road map with a timetable for specific goals, objectives and actions at the international level. It also promoted the importance of the "valuation of biodiversity and ecosystem services and the integration of these valuations into policies, decision-making and economic processes" (European Union, 2012c). The Union also continued to promote the adoption of the Ten-Year Framework of Programmes on Sustainable Consumption and Production throughout the Rio+20 preparatory process.

By the start of the conference, the European Union's position was no longer one of promoting a green economy road map. Rather, it recognized that "[t]he possibility now offers itself to work with a broad range of willing countries to develop policies on the green economy as a common undertaking" (European Union, 2012). The Union also supported the upgrading of UNEP into a WEO and a change in the level in the United Nations at which sustainable development would be addressed.

The greatest divergence within the European Union was between the development community, fixed on the MDGs, and the environmental community. This was resolved to some extent when it welcomed the proposals for

Sustainable Development Goals and supported the proposal that these goals should be universal and be linked to concrete targets and indicators. On April 19, 2012 the Union tabled a set of operational and time-bound targets in areas directly related to the transition towards an inclusive green economy which it hoped the negotiators would be prepared to negotiate with international partners to ensure an ambitious, action oriented outcome of the conference (European Union, 2012b). These were on:

- water;
- oceans and marine environment;
- sustainable land management and ecosystems;
- sustainable energy;
- resource efficiency.

As negotiations evolved it became clear that, even with the excellent work done by the EU team under Timo Mäkelä and Hugo Schally, more work needed to be done to draw a clear distinction between the proposed targets for a transition to an inclusive green economy and the definition of thematic areas, goals and targets for the Sustainable Development Goals for which European Union countries were not ready to agree before or during the conference. Although the EU withdrew its proposal of "transition targets," it scored a significant success by influencing in "The Future We Want" the vision for the Sustainable Development Goals. The European Union considered that these goals should be:

> coordinated and coherent with the Millennium Development Goals (MDGs) review process, without deviating efforts from the achievement of the MDGs by 2015; further considers that it would be important to have an overarching framework for post 2015 that encompasses the three dimensions of sustainable development with goals that address key challenges in a holistic and coherent way to ensure the optimal mix of measures for attaining lasting solutions.
>
> (European Union, 2012c)

In the area of the institutional framework for sustainable development, the European Union supported the upgrading of UNEP "into a specialized UN agency for the environment based in Nairobi with a revised and strengthened mandate and universal membership" (European Union, 2012c). It also supported the putting into place of a strong governance structure for sustainable development.

Immediately at the end of the last preparatory committee, the European Union was among the actors expressing surprise at the lack of detailed negotiation in Rio once Brazil took over the chairing of the negotiations on the Outcome Document. The Union was also somewhat discontented at the overall level of ambition, as it had hoped until the last minute that the level of ambition could be raised in certain areas, so overall saw Rio+20 as a starting point for action.

A joint press release on June 19, 2012 by the EU Commissioner for the Environment, Janez Potočnik, and the Danish Minister for the Environment, Ida Auken (the six-month rotating presidency of the Council of the European Union by Denmark was coming to an end on June 30, 2012), who was also current chair of the European Environment Council, commented as follows:

> The European Union in broad terms welcomes the Rio+20 declaration on the future we want, although a number of ambitions were not fully achieved. The EU has worked hard to secure a positive outcome on several fronts at Rio, including on the green economy, engagement in key areas for future action and on sustainable development goals. Some degree of institutional reform has also been achieved.
>
> In a joint declaration, the Danish Environment Minister and current chair of the European Environment Council, Ida Auken, and EU Environment Commissioner Janez Potočnik said:
>
> *"Today the world has indicated that we must move towards sustainable development and the inclusive green economy is a central pathway to achieve this. The EU has remained committed and constructive throughout the negotiations, and has spoken with one ambitious voice. We are pleased that this delivered results. However, we recognize that Rio is just the beginning and a range of activities have to be followed through at international level. The EU is committed to eradicate poverty and to improve people's lives while significantly reducing pressures on the environment."*
>
> Through persistent negotiation effort, the **green economy** is now understood as an important tool for achieving sustainable development. It will enhance our ability to manage natural resources sustainably, and it is recognised that urgent action to tackle unsustainable patterns of production and consumption has to be addressed wherever these occur. The possibility now offers itself to work with a broad range of willing countries to develop policies on the green economy as a common undertaking.
>
> The EU has also secured engagement in a number of **areas** such as water, oceans, land and ecosystems, food and sustainable consumption and production (SCP). In particular: a ten year framework for action on SCP has been adopted, the human right to water has been described, as well as a range of commitments such as on avoiding land degradation and on oceans, including on tackling marine litter. We are happy that the international community has now committed to take a decision on negotiations on an implementing agreement under UNCLOS [the United Nations Convention on the Law of the Sea] – an important step for the protection

of marine biodiversity. However the EU was not successful in getting agreements on specific targets in a range of other areas.

At the same time there has been agreement to launch a process in Rio on **Sustainable Development Goals**, primarily driven as an intergovernmental process. The EU is ready to fully engage in such an international process at all levels.

On **institutional reform** the EU did not reach all its ambitions. In its quest for establishing a World Environmental Organization, after long negotiations the EU managed to achieve a strengthening of a number of functions of the existing UN Environment Programme. In addition, to promote sustainable development governance a Forum has been established.

On **means of implementation**, agreement was reached to establish an intergovernmental process to develop options on a Sustainable Development Financing Strategy to mobilise financial resources. The EU as a whole remains committed to the Official Development Aid (ODA) level of 0,7% of GDP by 2015, and has been steadily increasing its ODA since the Johannesburg Conference in 2002.

Engaging with the non-governmental sector such as businesses and civil society is crucial for the green economy and sustainable development. The Rio outcome calls on business and industry to engage in responsible business practices and corporate sustainability.

(European Union, 2012a; original emphasis and italics)

### *Key EU negotiators*

Hugo Schally was one of the key negotiators for the European Union during the Rio+20 negotiations. He was a strong supporter of the Sustainable Development Goals and the green economy. The Union put forward robust positions on strengthening UNEP and supporting a Sustainable Development Council as a replacement for the CSD. Under his leadership the European Union was also very supportive of involving stakeholders through the preparatory process, a position that was translated into the Union's becoming a major funder for stakeholder engagement activities. Schally had also been the main EU negotiator of the Nagoya Protocol on the Access to Genetic Resources and the Fair and Equitable Sharing of Benefits Arising from their Utilization, adopted in 2010 under the Convention on Biological Diversity.

Ida Auken was Minister for the Environment, Forests and Nature for Denmark during the Rio+20 process, at a moment when Denmark was in charge of the six-month rotating presidency of the European Union. She promoted corporate social responsibility as one of the issues Denmark wanted to see advanced at Rio+20. Since then, Denmark has been one of four countries

*Figure 3.3* Ida Auken, Danish Minister of the Environment, intervening before the closing of the informal consultations led by the host country. She is joined by, from left to right, the Austrian ambassador to the United Nations, Thomas Mayr-Harting, EU Commissioner Janez Potočnik and the head of the European Commission's Environment Directorate-General, Karl Falkenberg (courtesy of IISD/*Earth Negotiations Bulletin* (www.iisd.ca)).

– with France, South Africa and Brazil – to take forward the work on a Global Agreement for Sustainability Reports to be produced by companies listed on the stock exchange. Auken is a Danish politician from the Socialist Party. Her uncle, Svend Auken, held a similar post in the Danish government in the 1990s and was instrumental in securing the World Summit on Sustainable Development in 2002.

Stephan Contius was the head of division at Germany's Federal Ministry for the Environment, Nature Conservation and Nuclear Safety. Contius was Germany's face for the positions it took and promoted in the Rio+20 process. On Germany's and the European Union's behalf he supported the concept of the green economy as a mechanism for integrating the three pillars of sustainable development to realize social and economic growth, and promoted the idea of a green economy road map and a United Nations Environment Organization. The two latter ideas were not embraced by member states and were not part of the Rio+20 outcome. That should not, however, define Contius's contribution to Rio+20, since he was also the promoter of one of the valuable and tangible outcomes of the Bonn 2011 Nexus Conference on the water–energy–food nexus, which fostered interlinkage thinking on these important and connected elements of development. Germany emphasized the location of these elements within the social, economic and environmental pillars of sustainable development as central to the green economy and green growth. Nexus thinking is intended not only to show the interconnections to various problems but also to bring about cross-sectoral solutions to solve multiple and connected problems. In this regard, Contius was very much a part of advancing one of the important ideas that emerged during the Rio+20 preparatory process.

Jean-Pierre Thébaut was France's ambassador for the environment and Rio+20. During the process, he led France's and the European Union's push to upgrade UNEP into a World Environmental Organization. France also played a significant role in three other central aspects of "The Future We Want": the agreement on oceans and seas; the strengthening of the Economic and Social Council; and the promotion of the water agenda, hosting (in Marseille from March 12 to 17, 2012, during the Rio+20 preparatory process) the sixth World Water Forum.

Kitty van der Heijden was the Netherlands' ambassador for sustainable development for Rio+20. Her responsibility was sustainable development issues, including water, sanitation and food security. Van der Heijden was fully engaged in the Rio+20 process and also gave support to the Natural Capital Declaration, which is a declaration by 37 CEOs of institutions in the wider financial sector, such as banks and insurance companies. It describes itself as "a global statement which demonstrates the commitment of the financial sector to work toward integrating natural capital criteria into financial services and products." Van der Heijden was also instrumental in collaborating with and supporting UN DESA on the hosting of the meeting "Realizing Inclusive and Green Growth: UN Rio+20 Business and Industry Consultation with Government and Civil Society," which took place on April 11 and 12, 2012 in The Hague (UN DESA, 2012). This meeting served to generate momentum, which would end in an unprecedented number of businesspeople attending Rio+20, where they considered the issue of business as a driver for sustainability, growth, job creation and development, while simultaneously building profitability and shareholder value.

### *Norway, Switzerland and Rio+20*

Norway and Switzerland defined constructive and to a large extent issue-based positions during the preparatory process of Rio+20. The fact that both European countries are not part of the European Union, allowed the flexibility to formulate specific and issue-based contributions, many of which have been identified so far in this book in areas such as sustainable energy, disaster risk reduction, gender equality, management of chemical products and wastes, sustainable forest management, mountains and green economy. In addition to these priorities, Norway and Switzerland played a supportive role in the outreach efforts by Colombia, Guatemala, Peru and United Arab Emirates for the Sustainable Development Goals. Both countries provided funding and political support through retreats and seminars for the SDGs, which were helpful in socializing the idea of one single set of goals beyond 2015.

Both Norway and Switzerland are strong supporters and major contributors to UNEP and placed, therefore, more emphasis in their negotiating approach to strengthening the functions, capacity and efficiency of UNEP and of the international environmental institutional framework. This approach was somehow contrasted with the strong advocacy efforts they displayed until the end of the preparatory process of Rio+20 for the establishment of a Sustainable Development Council as a new institution subsidiary to the General Assembly. After the adoption of "The Future We Want" they played a constructive role in setting the agreement for the High Level Political Forum under the auspices of ECOSOC to develop a review and implementation mechanism for the Post-2015 Development Agenda drawing from the experience of the Annual Ministerial Review.

## The United States of America and Rio+20

Like the European Union, the United States was reluctant to hold the proposed Rio+20 Conference. On September 29, 2009, coordinated by the National Resources Defense Council (NRDC), 13 of the top US NGO leaders jointly wrote to President Obama as follows:

> We are writing, on behalf of civil society organizations representing more than a million Americans, to request that the U.S. Government enthusiastically support the proposal now before the United Nations to hold an Earth Summit in Brazil in 2012. We hope that you will see the Summit as an opportunity to consolidate the gains made in your first Administration towards sustainable development and to catalyze actions worldwide to build a new green global prosperity.
>
> (NRDC, 2012)

From the very beginning, the United States lobbied hard for a small number of days for the preparatory process for Rio+20 and an outcome that would be a short five-page political declaration which its leaders hoped would work to inspire the mainstream. This was in response to the failure at Copenhagen to secure a new climate agreement. The United States had a clear view that after Copenhagen, Rio+20 was a meeting that was "too big to fail." It also saw Rio+20 as an opportunity to move away from traditional negotiated text to focus on all stakeholders making commitments. This echoed President Bush's position at the WSSD in 2002.

Yet the arrival of President Barack Obama at the White House signalled a renewed commitment towards sustainable development, the green economy and the multilateral system. In this regard, Rio+20 was increasingly perceived as an additional opportunity for enhancing those objectives at home and abroad, in a manner consistent with the United States' new "Global Development Policy." This policy attempted to "elevate development as a central pillar of [its] national security policy, equal to diplomacy and defense" (White House, 2010). The three main components of the strategy included three voluntary initiatives on climate change, health and food security.

In its submission to the Zero Draft (United States, 2011), the United States focused on support for the green economy and three key messages:

- the built environment: clean energy and urbanization;
- the natural environment: ecosystem management and rural development;
- the institutional environment: modernizing global cooperation.

Throughout Rio+20, the position of the United States on several institutional issues, such as the transformation of UNEP, was clearly defined. While the United States did not support the upgrading of UNEP into a WEO, it did engage constructively in defining incremental measures to address some of the main bottlenecks affecting UNEP and the environmental pillar, such as reforming UNEP's mandate, promoting universal membership of its Governing Council, strengthening its operational and monitoring capacities, and reforming the Committee of Permanent Representatives. Nevertheless, the reluctance of the United States to engage on the transformation of UNEP into a specialized agency often led to polarized debates between the United States and the European Union reminiscent of those during the Copenhagen UNFCCC meeting, raising concerns that the leadership had definitely moved away from the rest of the developed world in the area of sustainable development.

The approach of the United States solidified around the hope that having some agreed targets and goals would result in a bottom-up approach whereby stakeholders and citizens would help achieve these goals and targets. As Reta Jo Lewis, special representative for global intergovernmental affairs of the US government, said:

> Rio+20 is about the future. The United States believes that Rio+20 should be a different kind of meeting, one that transforms the multilateral approach to sustainable development and incorporates its concepts across all sectors. It is our hope that Rio+20 will be truly inclusive of a broad collection of stakeholders, including state and local officials, civil society and the private sector.
>
> (Lewis, 2012)

Even with a bottom-up approach there would need to be much more effective reporting and monitoring of companies' activities. The major proposal on the table of a global reporting framework would have gone a long way to ensuring that companies would be held accountable in the market for their environmental activities. Unfortunately, this was not supported by the United States, despite being a central recommendation of the Global Sustainability Panel's report.

Among the contributions made by the United States to "The Future We Want" can be highlighted a greater recognition of the role of science and technology. Information, particularly data on economic, social and environmental issues, and science were deemed critical components to guide policy decisions and inform development planning. In addition to this, the United States was supportive of voluntary initiatives launched by the Secretary-General, such as the United Nations Sustainable Energy for All; by the UN Foundation, such as the Global Alliance for Clean Cookstoves, which aims to help 100 million families adopt clean cooking stoves and fuels by 2020; and finally, by coalitions of countries, as was the case with the Climate and Clean Air Coalition, whose goal is to reduce short-lived climate pollutants that cause more than 30 per cent of current global warming.

### *Key US negotiators*

Ambassador Elizabeth M. Cousens was sworn in on April 2, 2012 as US representative on the United Nations Economic and Social Council and alternate representative to the UNGA. Ambassador Cousens has served as Sherpa to Ambassador Rice for the UN Secretary-General's High-Level Panel on Global Sustainability and played an important role in the final run-up to Rio+20 and in the preparation of the Post-2015 Development Agenda.

Todd Stern is the special envoy for climate change at the State Department. He joined the US Rio+20 team at the final New York PrepCom, representing a higher engagement of the White House in Rio+20. He focused on section V of the Rio+20 negotiations. He plays a central role in developing US international policy on climate and is the President's chief climate negotiator, representing the United States internationally at ministerial level in all bilateral and multilateral negotiations regarding climate change.

John M. Matuszak was division chief for Sustainable Development in the US State Department. In the final part of the negotiations he focused on the green economy, and particularly focused on the inclusion of natural capital in the Rio+20 text.

Rebecca Weber was the lead officer coordinating US engagement with the United Nations on all aspects of sustainable development policy; she was particularly relevant in pursuing her country's engagement with regard to the strengthening and upgrading of UNEP and the definition of the High-Level Political Forum.

## A final push to conclude the process

The last meeting of the preparatory committee took place in Rio on June 13–15. "We are near, and yet so far," said UNCSD secretary-general Sha Zukang in his closing remarks (IISD, 2012a).

For Brazil's incoming presidency of the conference, the crucial moment its team had been quietly preparing for was near. In this section we will highlight some of the contributions made by Brazil to the overall narrative of the Rio+20 Conference, specifically after the new government of President Dilma Rousseff assumed the presidency. We will also look into the preparatory work that Brazil undertook, mainly through organizing informal consultations during 2011 and 2012 which had the purpose of helping the Brazilian authorities understand the role that countries were expecting them to play in ensuring the success of the conference. Finally, we will chronicle the role played by Ambassador Luiz Alberto Figueiredo Machado (who became Brazil's Foreign Minister on 26 August 2013) and his team in steering the transition from the third meeting of the preparatory committee towards closing the gaps in the text of "The Future We Want."

### *A new political discourse*

The role of Brazil as host country went beyond the offer to host a celebratory conference to mark two decades after the 1992 Earth Summit. A first phase of the Brazilian leadership involved the original conceptualization, between 2006 and 2009, of why, 40 years after the 1972 Stockholm Conference on the Human Environment and 20 after the 1992 Conference on Environment and Development there was a need for a follow-up conference on sustainable development. The original idea behind the conference and the political drive to host it were illustrated by the 2007 speech of President Lula at the opening of the 2008 session of the UNGA (covered in Chapter 1).

The election in October 2010 of Brazilian President Dilma Rousseff, who took office on January 1, 2011 in succession to her friend and mentor President Lula, happened just over a year after the official organizational arrangements, themes and objectives for Rio+20 had been agreed. For a while, most observers assumed that there was going to be continuity in the political discourse towards Rio. However, after several months a new political discourse from the Brazilian authorities began to emerge.

The Brazilian leadership in fact reassessed the original narrative of the conference. In a nutshell, the emerging vision of Brazil for Rio+20 before 2011 was "twenty years after Rio"; afterwards, it was oriented to the challenges for "the next twenty years from Rio." This vision – which ultimately facilitated a drive for a converging and transformative global development agenda – had two main components. The first one, as has been said, was a shift from the original narrative oriented towards looking at the gaps of the past two decades in the implementation of Agenda 21 to one of an opportunity for the international community to look forward to the challenges lying ahead in the forthcoming 20 years. Such a simple, yet powerful, idea was grounded in the political discourse of President Rousseff, who advocated that her country, Brazil, had to keep on changing. A second shift provided by the new Brazilian leadership was a strong advocacy for making the fight against poverty and international inequalities a core component of the sustainable development agenda. Although poverty had always been there as a central component of the Rio 1992 and Johannesburg 2002 dynamics and outcomes, the new Brazilian narrative for the conference made it possible – in the context of the debate forced by the Sustainable Development Goals – to stir into the global discourse the idea of a transformative development agenda. In this agenda the eradication of poverty and the fulfilment of basic human needs worldwide was "only the beginning."

Brazil also promoted greater transparency and involvement in the Rio+20 Conference of NGOs and other stakeholders, particularly from academia. The initiative of the Brazilian authorities in convening the Rio Dialogue Days, held back to back to the leaders' summit, as well as in their preparatory global online consultation, speaks to their concern to give an opportunity to participate for many actors that do not necessarily have access to the UN negotiating processes. These dialogues, all constructed with the help of the United Nations Development Programme (UNDP), pointed towards the need for a new development agenda in which new, universally applicable priorities would emerge, such as sustainable consumption and production, as a complement to those already covered by the MDGs. The Rio Dialogue Days represented in addition not only a first chance for civil society to assess and question the achievements and limitations of "The Future We Want" but also the first opportunity – and indeed the blueprint – for the global conversation on the development agenda beyond 2015 (consisting of national consultations, a global web-based platform and thematic issue-based consultations), kick-started by the UN system shortly after Rio.

The third element brought forward by Brazil was to use Rio+20 as an opportunity to highlight the growing importance and emerging role of developing countries. The economic trends pointed out in Chapter 1 reflect the increasing political, economic and technological importance of developing countries. This importance invited, from the perspective of key Brazilian intellectuals and policy-makers, the reassessment and reinvention for the next decades of the international cooperation paradigm without compromising the different responsibilities of the North and the South respectively.

This new vision in no way represented a rupture with the past. It maintained the importance of discouraging the idea that Rio+20 should be associated exclusively with "environmental issues" and also retained the strong focus on inequalities, essentially perceived at the international level between advanced and developing countries, perhaps missing the opportunity to broaden Lula's original discourse towards addressing inequalities at all levels.

### *The importance of informal consultations*

Throughout the process leading to Rio+20, Brazil organized a series of parallel informal consultations in order to explore potential avenues for progress in areas that had emerged as controversial during the formal preparatory process. The format and methods of these meetings borrowed from a model of informal consultation developed by Mexico in 2009 and 2010 during the run-up to the successful Sixteenth Conference of the Parties and Sixth Meeting of the Parties of the UN Framework Convention on Climate Change and the Kyoto Protocol, where the Cancún Agreements were adopted, breaking the negative spell cast by 2009.

The Brazilian informal consultations attempted to bring around the same table a representative group of key delegations for open and frank discussions with no planned outcome. Ambassador Luiz Alberto Figueiredo Machado, Brazil's Undersecretary for the Environment, Energy, Science and Technology at the Ministry of External Relations, always led these informal consultations. He was joined at all times by Ambassador André Corrêa do Lago, Brazil's chief negotiator for Rio+20.

The first of these consultations was held in Rio de Janeiro as early as August 2011. It was limited to some 40 governmental delegations and included the participation of members of the UN Secretary-General's High-Level Sustainability Panel, as well as one of the executive coordinators for Rio+20, namely Brice Lalonde. Among the many issues discussed, there were some emerging themes that would prevail throughout the process, such as the message that the eradication of poverty remained the first priority for most countries. It was also found that there was broad dissatisfaction regarding the lack of coordination and coherence of the functioning of the UN system in the field of sustainable development and the lack of political will to transform the CSD, whose performance had not met the expectations with which it had been created. Third, there was also a strong call to strengthen the environmental pillar (e.g. by enhancing UNEP and ensuring universal membership), but without altering the paradigm of sustainable development or weakening the balance between the three pillars. Fourth, right from the first informal consultation the link between sustainable consumption and production and the green economy emerged as an area where most progress could be made. Finally, there was a common understanding on the need to strengthen stakeholder participation in the conference and in future institutional arrangements, and an overall concern that the international financial institutions, as well as finance ministries, had so far remained disengaged from the process.

A second consultation, with Latin American and Caribbean delegations, was held in Brasilia on March 13–14, 2012 and sought to gather regional views and perspectives that could result in a more unified contribution from the region to the Rio+20 process.

The third and last consultation was held in Brasilia on April 12–13. It had a greater sense of urgency and tension, given the impasse reached on several of the major themes after most of the preparatory process had run its course. Forty-five delegations from all regions represented at the level of vice-ministers and ambassadors to the United Nations convened at the majestic Palace of Itamaraty. There was a greater level of political convergence in this last set of informal consultations than in the expert-led negotiation in New York thus far. On this occasion, Brazil had the opportunity to reassure participants about its approach and the results it expected as host and incoming president of the conference. It also became clear that the conference host country had managed to build strong support from developing countries and key developed countries. There was broad positive agreement on the need to integrate the social, economic and environmental dimensions of sustainable development, as well as on the benefits of defining a set of Sustainable Development Goals that reflected this integration.

On the green economy it became clear, as some Latin American countries stated, that Rio+20 would build its own approach to it as one of the means to achieve sustainable development. There were also still open divisions regarding institutional arrangements, with fewer delegations still favouring the creation of a Sustainable Development Council and the establishment of a specialized agency that would replace UNEP. Regarding the latter, there was a contrasting sense that the financial situation the world was facing did not allow for such a level of ambition. This last point was reported in the Brazilian newspaper *O Globo* on April 14 as follows: "Brazilian negotiators did not wish Rio+20 to be marked as an environmental conference, as was the case for Rio '92." For this reason, according to *O Globo*, delegations meeting at Itamaraty concluded that "agreements must be reached on sustainability goals that consider all three pillars of development, a social-environmental protection floor for the eradication of poverty, and the accounting of the environmental component within the GDP" (Carvalho, 2012).

The final track of informal dialogue was the conversations convened by Ambassador Luiz Alberto Figueiredo Machado with a representative yet focused group of heads of delegations on the issue of the strengthening and upgrading of UNEP. From those conversations emerged a non-paper which ultimately became the compromise proposal discussed in the pre-conference informal consultations and ultimately reflected in paragraph 88 of "The Future We Want." (A non-paper is a proposed agreement or negotiating text circulated informally among delegations for discussion without committing the originating delegation's country to the content.)

### The pre-conference informal consultations

On June 12 the Rio+20 Bureau met at Rio Centro, the main venue of the conference, for a planning and preparatory session on the eve of the start of negotiations. When negotiations opened the next day, the approach of two large working groups with some 11 splinter groups continued. By the end of the first day there had been no significant movement on the text. Positions inherited from New York negotiation sessions were by then entrenched. When the Bureau met on June 14, the mood was one of concern. Some negotiating groups had made no progress, while others were in a state of high agitation. There was intractability in the positions of some groups, particularly within the Group of 77 on means of implementation (MOI) and of the European Union on the green economy road map with targets.

The institutional discussion continued in a state of weariness after delegates had decided, in early June, to base their discussions on a non-exhaustive list of tentative functions that the High-Level Political Forum could perform, without addressing any details concerning its institutional form. The chair of the institutional splinter group had also decided to avoid the discussion on international environmental governance (IEG) in order to focus on the Forum's functions. This lack of balance between the two subjects frustrated many negotiators, who had less than three days to build, on the basis of the Group of 77 compromise text, a package of pragmatic measures leading to the enhancement of UNEP, building on many years of IEG discussions. The Bureau was informed that the Group of 77 was threatening to pull out of the negotiations if the developed countries did not make concessions on MOI, particularly on the establishment of a technology facilitation mechanism. The words "arresting contagion," "deadlock" and "crisis aversion" were being used in the corridors.

Co-chair Ambassador Ashe insisted that in order to make progress, the negotiations would have to continue through the night. The Bureau agreed to get a sense of what was happening in the splinter groups and reconvene at 10 p.m. On adjournment, the co-chairs were to assess at first hand what was really happening in the splinter groups and how much damage control was required to prevent a collapse of the negotiations.

Within a short time, word had spread across the conference compound that the negotiations in the MOI splinter group had completely broken down and that the Group of 77 was pulling out of the negotiations. On investigation, however, it was found that tensions and verbal exchanges had run very high in the rift between the Group of 77, on the one hand, and the European Union, Canada and the United States, on the other. Contrary to reports of breakdown and withdrawal of the Group of 77 from the negotiations, facilitator Selwin Hart, a very experienced and committed negotiator from Barbados, was drawing on his full suite of diplomatic skills and experience to expertly shepherd the Group away from crisis mode and into making some progress.

*Figure 3.4* Selwin Hart (centre), the very experienced and committed negotiator from Barbados, during the Third Preparatory Committee (courtesy of IISD/*Earth Negotiations Bulletin* (www.iisd.ca)).

By the following day, June 15, a conciliatory mood had been restored. Hart had offered his splinter group a facilitator's text. He succeeded in breaking the deadlock and getting the Group to reach consensus on the more difficult issues on which there had been an impasse. This positive movement helped in achieving similar last-minute successes in other negotiating groups.

On that morning, the final day of the preparatory committee, and after an evening of high tension and pessimism, the clouds lifted and optimism broke through as the talks continued without collapsing. The Bureau and the co-chairs were united on the strategy to deploy for the remaining time: no paragraphs were to be reopened; there was to be an all-out push to get an agreed text. Co-chair Ambassador Kim would oversee three splinter groups; co-chair Ambassador Ashe, eight. It was also established that Brazil would take over the management of the negotiations of the working and splinter groups from the end of the next day, June 16, until June 20, when the high-level segment of the conference would start. It was agreed that a closing ceremony for the preparatory committee would take place at 10 p.m. on June 16 but negotiations would continue after the ceremony, until 11 p.m. The devoted UN Secretariat would once more work through the night, updating the text to reflect the day's agreements; at that moment, the text as it then stood would be physically handed over to Brazil.

At 11.15 p.m. on June 16, the closing ceremony for the preparatory committee took place. It began with the facilitators giving updates on the status of the negotiations of the splinter groups, all of which were able to report good progress. The meeting then became more formal and a number of agenda items were addressed. In his concluding remarks, co-chair Ambassador Ashe noted that, while good progress had been made, there were still some thorny issues keeping member states apart. He reminded participants that the co-chairs could play no further role since their "life" and that of the Bureau were now coming to an end.

The minister of foreign affairs of Brazil, Antonio Patriota, thanked the UN Secretariat and the co-chairs for their efforts and guidance, and expressed appreciation to the facilitators of the splinter groups. And with that, Brazil took over the chairmanship of the negotiations and the presidency of the conference. The minister indicated that Brazil would receive the text the next morning, conduct informal consultations and "hear all voices," then the meeting would be reconvened at midday to outline Brazil's proposal on how the work should continue. At 12:15 a.m. that morning, proceedings came to a close, with Brazil now firmly chairing.

When Brazil took over, time was of the essence. Heads of state and government would be seated in the plenary hall at 10 a.m. on the morning of Monday June 20 and a political document of substance had to be ready for them to sign. There was a real risk that, given the controversy, the protracted negotiations and the predictions for a less than successful outcome, without an agreed text the leaders might not turn up and the conference could end in genuine failure. It was therefore very apparent why Brazil as host and chair of the conference was paying such close attention to the negotiations. Starting from the second round of informal negotiations held in April, Brazilian representatives sat near Group of 77 negotiating coordinators in all working groups and splinter groups and were seen to be very closely monitoring progress in negotiations and positions by all member states. The key at this time was to get the text to an outcome that would meet the needs of all interest groups, addressing the more controversial and divisive issues without weakening the text to the extent that it would be ineffectual.

The following afternoon, June 17, Brazil, after consultations with negotiators, disseminated its compromise text. The text that delegates received on that afternoon was one that each of them knew practically by heart. It was not a "new" document. It contained no last-minute surprises. The Brazilian text was the co-chairs' document that had undergone months of negotiation. It remained faithful to the wish expressed since March of retaining country ownership of the proposals. The experienced Brazilian negotiators, who had followed the process closely, had closed the remaining gaps with balanced references drafted taking into account their own negotiating experience, the general thrust of the discussions, the agreements made in the splinter groups and the intimate knowledge of the breaking points within the Group of 77. It also reflected the findings derived from the informal consultations with member states. A few hours were given for delegations to review the text. Quiet optimism and poker faces now pervaded the meeting halls.

With the screens displaying the text in a negotiation room now literally shut down at last, Brazilian minister Patriota and Ambassador Figueiredo invited delegations to start afresh – not a drafting or a persuasion exercise, but a real last-minute negotiation with the purpose of allowing delegations to make clear where their red lines had been crossed and all participants to close final agreements in a transparent manner. Five working groups were outlined. Each one of them was presided over by some of the most talented and experienced Brazilian diplomats from the multilateral environmental and sustainable development negotiations, selected by Ambassador Figueiredo and Ambassador André Corrêa do Lago. Among them were Maria Teresa Pessoa, Raphael Azeredo, Paulino Franco de Carvalho Neto, Fernando Coimbra, Angelika Ikeda, Antonio di Ricardi, Paulo Chiarelli and Pedro Fiorencio Cabral de Andrade. This approach worked.

At 10.30 a.m. on June 19, the day before the high-level segment was due to start, exhausted ambassadors, negotiators, stakeholders, a relieved UN staff and a justifiably proud and happy Brazil convened. Ambassador Figueiredo, who had been responsible for the conference on Brazil's behalf, along with the officials from the Ministry of Foreign Affairs, and had brought the sharpness of his legal mind to directing the revised text and consultations, sat next to Foreign Minister Patriota. A beaming conference secretary-general Sha Zukang shared the podium with them. Minister Patriota spoke, indicating that the plenary "would adopt the text *ad referendum*." Before he could bring down the gavel, Kenya and the Republic of Congo on behalf of the African Union spoke on a point of order. There followed a two-hour deluge of speeches by member states expressing thanks to Ambassadors Ashe and Kim, Sha Zukang, Secretariat

*Figure 3.5* A view of the dais after the adoption of the text during the closing plenary of the pre-conference informal consultations. Pictured are Brazilian Foreign Minister Antonio de Aguiar Patriota (centre), Ambassador Luiz Alberto Figueiredo, André Corrêa do Lago, Ambassador Maria Luiza Viotti and team (courtesy of IISD/ *Earth Negotiations Bulletin* (www.iisd.ca)).

officers and the Brazilian hosts for their efforts in achieving consensus on an outcome document for leaders to sign. Many speakers also voiced concern about elements of the outcome document and what they perceived as inclusions and omissions with which they did not agree.

Two factors contributed to acquiescence to the text: the fear that opening even one paragraph was dangerous and would create a precedent that would see the reopening of the entire text by some discontented party, under the negotiating practice that "nothing is agreed until everything is agreed"; and a genuine desire to have a resolution and outcome document. Sha Zukang thanked everyone for their work and gave the first synopsis of exactly what had been achieved, and the new ground broken in the Rio+20 Outcome Document. At 12.15 p.m., Minister Patriota brought the gavel down on two and a half years of work and a highly charged final week of negotiations, to thunderous applause.

## No perfect outcome: the gaps in "The Future We Want"

There is no such thing as a perfect outcome in a multilateral negotiation, particularly when decisions are adopted without recourse to a vote. In order to evaluate the gaps within "The Future We Want," it is necessary to understand what countries were actually willing to commit to. During the pre-conference informal consultations led by the host country, transparency was ensured. Meetings were open to all delegations and detailed records of each session were published and available online through the excellent International Institute for

Sustainable Development's *Earth Negotiations Bulletin*. Through these records, it is possible to identify some of the most glaring gaps:

- the application of the principle of common but differentiated responsibilities and respective capacities;
- elaboration of Sustainable Development Goals or, at least, agreement on priority areas;
- development of a proper narrative for implementation for the thematic areas;
- the guarantee of specific mechanisms to deliver financing and technology transfer;
- a road map to address subsidies and trade in environmental goods and services;
- broader measures of progress and a robust registry of commitments and how they would be enforced;
- the strengthening of ECOSOC;
- the need to recognize and operationalize intergenerational solidarity for the achievement of sustainable development, taking into account the needs of future generations;
- the transformation of UNEP into a specialized agency;
- a requirement for companies listed on stock exchanges to produce corporate sustainability reporting;
- the absence of a reference to reproductive rights;
- a lack of ambition on various sectors of the green economy and the elimination of time-frames, goals and targets;
- lack of a menu to address the financing of sustainable development and a failure to highlight LDCs, both under the area of MOI.

It is hoped that these gaps will be addressed in the process for developing new development goals in 2015.

In terms of the process followed, perhaps there are some lessons that might be worth considering for the next conference:

- appoint an independent secretary-general for the conference who can commit all his or her time to the conference;
- set up an independent secretariat that can draw staff from different UN agencies and programmes, governments and stakeholders;
- allow expert working groups to develop a first-level text for governments to comment on;
- go back to proper preparatory processes with preparatory committees of two weeks, and thus address the inextricable issue of what resources to raise and dedicate to preparatory meetings;
- host the preparatory committees not only in New York but also in other UN centres (e.g. Nairobi and Geneva.);
- never use screens to project text for negotiating when the nature of negotiations is political and the size of the group is not appropriate for the use of screens.

## References

Ahmia, M. (ed.) (2006) *The Group of 77 at the United Nations: The Collected Documents of the Group of 77*, New York, Oxford University Press. Available online at: http://mouradahmia.g77.org/volume1.php

Carvalho, E. (2012) Behind Closed Doors, Countries Try to Unlock the Rio+20 Negotiations, *O Globo*, April 14. Available online at: http://g1.globo.com/natureza/noticia/2012/04/portas-fechadas-paises-tentam-destravar-negociacao-da-rio20.html

Chasek, P. S. and Wagner, L. M. (2012) *The Roads from Rio: Lessons Learned from Twenty Years of Multilateral Environmental Negotiations*, New York, RFF Press

European Union (2011) Rio+20: Towards the Green Economy and Better Governance, Brussels, EU. Available online at: http://eur-lex.europa.eu/LexUriServ/LexUriServ.do?uri=CELEX:52011DC0363:EN:NOT

European Union (2012a) A Joint Press Release Giving Initial Views from Commissioner Potočnik and Ida Auken, Danish Environment [*sic*] for the Environment, June 19, Brussels, EU. Available online at: http://ec.europa.eu/environment/international_issues/pdf/rio/report_19_06_2012.pdf

European Union (2012b) Commission Non-Paper Rio+20 for the Informal Environmental Council, April 19, Brussels, EU. Available online at: www.eurostep.org/wcm/dmdocuments/Informal%20Meeting%20of%20Ministers%20for%20Environment%20in%20Horsens%2018%20%2019%20April%202012%20-%20Presidency%20conclusions.pdf

European Union (2012c) Rio+20: Pathways to a Sustainable Future – Council Conclusions, March 9, Brussels, EU. Available online at: www.consilium.europa.eu/uedocs/cms_data/docs/pressdata/en/envir/128881.pdf

G77 (2008) G77 Ministerial Declaration, September 28, New York, G77. Available online at: www.g77.org/doc/Declaration2008.htm

IISD (2012a) Summary of the Third Round of UNCSD Informal Consultations, May 29–June 2, New York, IISD. Available online at: www.iisd.ca/vol27/enb2740e.html

IISD (2012b) UNCSD Informal Consultations: Thursday, 3 May 2012, New York, IISD. Available online at: http://enb.iisd.mobi/uncsd-iinzod2/reports/enb2734e/

Lewis, R. J. (2012) The Road to Rio+20: International Goals and Subnational Actions, June 15, Washington, DC, Department of State. Available online at: www.state.gov/s/srgia/2012/194140.htm

NRDC (2009) Letter to President Barack Obama, September 29, Washington, DC, NRDC. Available online at: www.stakeholderforum.org/fileadmin/files/Earth_Summit_2012/Letter_to_President_Obama.pdf

Sarkozy, N. (2009) Speech in Brazil, London, Stakeholder Forum. Available online at: http://earthsummit2012.blogspot.com/2009/09/french-president-endorses-rio20.html

UN DESA (2013) Realising Inclusive and Green Growth: UN Rio+20 Business and Industry Consultation with Government and Civil Society Available online at: www.uncsd2012.org/index.php?page=view&nr=615&type=13&menu=23

United Nations (2012) Co-chairs' suggested text, paragraph 105, March 28, New York, UN.

United States (2011) Sustainable Development for the Next Twenty Years: United States Views on Rio+20, Washington, DC. Available online at: www.uncsd2012.org/index.php?page=view&type=510&nr=370&menu=115

White House (2010) Fact Sheet: U.S. Global Development Policy, September 22, Washington, DC, The White House. Available online at: www.whitehouse.gov/the-press-office/2010/09/22/fact-sheet-us-global-development-policy

Chapter 4

# Multiplying commitments

## Introduction

Ambassador Camillo Gonsalves of Saint Vincent and the Grenadines, speaking at the opening of the informal-informal negotiations, had challenged member states to move away from negotiating around language, brackets and punctuation to consider the strategic priorities and the critical question of what they would like Rio+20 to deliver, and what would be the key drivers that would make a difference to sustainable development globally and nationally.

Rio+20 was not only about process and negotiating dynamics; it was also about results and deliverables. We will highlight the major outcomes launched at Rio+20 that complement, and to a certain extent put in perspective, the official text and limitations of the agreements contained in "The Future We Want." Rio+20 mobilized all kinds of actors into delivering concrete practical results. This wider engagement started with the call by the co-chairs to all citizens of the planet to submit their input on what Rio+20 should deliver. These stakeholders put forth around 80 per cent of the 6,000 pages of initial proposals towards the Zero Draft of the outcome document.

It was followed by the Rio Dialogues, which constituted the preface to the movement for including all stakeholders in the effort of building a new development agenda. As an additional element on top of the Major Groups' efforts to convey the message of their constituencies, the Rio Dialogues did set a precedent that should be replicated in an improved manner in future sustainable development processes.

## The stakeholder Dialogue Days

Brazil initiated the Sustainable Development Dialogues with the support of the United Nations. They were held between June 16 and June 19 prior to the high-level segment of Rio+20. This set of ten Sustainable Development Dialogues brought together representatives from the nine Major Groups, as well as other stakeholders such as foundations, academics, philanthropic organizations and parliamentarians. In many ways these Dialogues built on a similar experience undertaken at the UN Habitat II Conference in 1996, while building new ICT capacity. The ten topics for the Sustainable Development Dialogues of Rio+20 were chosen by the Brazilian government in the last few months before Rio+20, in consultation with stakeholders. Prior to Rio+20, the UNDP launched a set of online platforms in order to provide the wider public with a democratic space for discussion on the ten themes of the Dialogues. Researchers from renowned academic institutions around the world facilitated the online debates on each of the ten. The Dialogues resulted in concrete recommendations for each area that were viewed and voted for on a public website. This was an innovative and more inclusive attempt to build a bridge between stakeholders and heads of state and government. Although the Dialogues came too late to have an impact on the negotiations, the fact that the top three recommendations from each of them were presented at the four High-Level Round Tables gave world leaders an idea of the priorities that they should take into account in the development of a new set of goals to build and replace the MDGs by 2015. Some governments believed that the outcome of the Dialogues could constitute an alternative for a short political document to be agreed should negotiations fail by the end of the third preparatory committee. The ten Dialogues and their top-three recommendations. abridged for readers' benefit, were as follows:

### *Unemployment, decent work and migration*

1 Put education at the core of the Sustainable Development Goals agenda.
2 Commit to a "decent work for all" goal and a "social protection for all" goal by 2030.
3 Compel national governments to respect the human rights of all migrant workers and their families.

### *Sustainable development as an answer to the economic and financial crises*

1 Promote tax reforms that encourage environmental protection and benefit the poor.
2 Create a tax on international financial transactions with a view to contributing to the Green Climate Fund.
3 Adopt shared Sustainable Development Goals (SDGs) that will be embraced by business, civil society and the public sector. These goals will include innovative metrics, public disclosure, public awareness and education at all levels.

### Box 4.1 Critical issues identified during the Rio Dialogues for any Sustainable Development Goals

- Empowering every place in the world to ensure that every individual meets their basic needs of health, safe water, sanitation and decent, dignified lives.
- Effective and efficient energy systems.
- A sustainable food supply, locally and globally.
- Sustainable urban environments, including initiatives on water systems, sewerage and other smart infrastructure.
- A committed industry required to clean up after itself.
- A shared set of principles and methods that will apply for all. This includes taxes, where we all see a very strong possibility for steering in the right direction; financial reform, including innovative financing for green development; and a shift to sustainable technologies in general.

#### *Sustainable development for fighting poverty*

1 Promote global education to eradicate poverty and to achieve sustainable development.
2 Ensure universal health coverage to achieve sustainable development.
3 Promote equitable access to information, participation, representation and justice at all levels.

#### *The economics of sustainable development, including sustainable patterns of production and consumption*

1 Phase out harmful subsidies and develop green tax schemes.
2 Include environmental damages in calculations of gross national product (GNP) and complement the latter with measures of social development.
3 Promote sustainable public procurement worldwide.

#### *Forests*

1 Restore 150 million hectares of deforested and degraded lands by 2020.
2 Promote science, technology, innovation and traditional knowledge in order to face forests' main challenge: how to turn them into productive resources without destroying them.
3 Zero net deforestation by 2020, respecting the rights and knowledge of peoples living in and from the forests and responding to their sustainable development needs.

### *Food and nutrition security*

1 Promote food systems that are sustainable and contribute to improving health.
2 Develop policies directed to both producers and consumers to encourage sustainable production of food supplies.
3 Eliminate misery and poverty-related malnutrition.

### *Sustainable energy for all*

1 Take concrete steps to eliminate fossil fuel subsidies.
2 Establish ambitious targets for moving towards renewable energy.
3 Scale up investments and political will to ensure the availability of universal, equitable and affordable access to sustainable energy services by the next decade.

### *Water*

1 Secure water supply by protecting biodiversity, ecosystems and water sources.
2 Implement the right to water.
3 Assert the importance of integrated water, sanitation, energy and land-use planning.

### *Sustainable cities and innovation*

1 Promote the use of waste as a renewable energy source in urban environments.
2 Plan in advance for sustainability and quality of life in cities.
3 Each head of state should identify a sustainable city to develop a network for knowledge-sharing and innovation.

### *Oceans*

1 Avoid ocean pollution by plastics through education and community collaboration.
2 Launch a global agreement to save high-seas marine biodiversity.
3 Take immediate action to develop a global network of international marine protected areas, while fostering ecosystem-based fisheries management, with special consideration for small-scale fishing interests.

## Voluntary commitments

The second innovation after the Dialogue Days was the introduction of a Registry of Voluntary Commitments (UN DESA, 2012b). The registry was intended to be a platform of voluntary commitments by all governmental and non-governmental stakeholders, aimed at poverty eradication and sustainable development. The registry, the first of its kind in the multilateral arena, employed an online registration format on which the pledger was to indicate the nature of the commitment, the resources to be committed by the pledging entity, the mechanisms against which the level of implementation and success would be measured, and the timeline for fulfilment. Commitments were on an entirely voluntary basis and could be on the thematic area of the stakeholder or pledger's choice. UN DESA gave directions only for the format of pledging (ibid.).

The addition of voluntary commitments made by all stakeholders to the sustainable development agenda was an interesting and beneficial mechanism that could secure a new level of consideration and action by stakeholders. The commitments represent a "bottom-up," self-policing and peer review approach to agreeing to specific actions for decreasing and eradicating poverty, spurring business action, advancing new policy and financing development activities, while simultaneously enlarging stakeholder engagement. The voluntary commitments constitute a key legacy of Rio+20.

The Natural Resources Defense Council noted that, over the years, although many commitments had been made in multilateral fora, there was no "universal mechanism for recording these pledges and tracking their progress" (NRDC, 2012). Céline Ramstein (2012) said that the commitments "offer a chance to further advance a sustainable future, with all the actors willing" to make them.

Another NGO commenting favourably on the Voluntary Commitments was the World Resources Institute (WRI):

> The Compendium, in a nutshell, would be a platform or registry of pledges by governments, civil society and the private sector that aims to fulfill globally-agreed-upon goals around a number of key themes, whether from prior (ratified but unfulfilled) multilateral environmental agreements, global sustainability initiatives, or principles enshrined in the 1992 Rio Declaration.
> (Foti, 2012)

The Registry attracted over 737 commitments from multiple actors, including 347 from member states, 98 from UN agencies and IGOs, and 591 from NGOs and other stakeholders. A further 300 were registered after the conference. The voluntary commitments were on almost all themes of Rio+20: water, oceans and seas, energy, the Higher Education Sustainability Initiative, green jobs and social inclusion, natural disasters and preparedness, sustainable consumption and production, food security, forests, sustainable transport, education, climate change, gender equality and public education on sustainable development. By the time the conference was over, the value of the Rio+20 Registry of Commitments had reached a total of US $153 billion (UN DESA, 2012b).

The Higher Education Sustainability Initiative, with nearly 40 per cent of the total commitments made, or just under 300, dominated the pledges. The second largest number of commitments came from the principles of the UN Secretary-General's Sustainable Energy for All initiative. The third area that saw a substantial body of commitments by 21 island states, on energy and sustainable development, emerged from the Small Island Development States (SIDS) Launch Conference of the Sustainable Energy for All initiative in May 2012 in Barbados (SIDS Energy for All, 2012). The Barbados meeting benefited from very high-level representation, with prime ministers, deputy prime ministers, ministers and ambassadors, who were all able to make decisions and commit their governments without reference to their capitals. The hosts of prior preparatory meetings had been unable to get delegates to make definitive decisions, since this power rested with capitals, to which delegates had to refer issues.

The NRDC described the Registry of Voluntary Commitments as "extraordinarily well received." Stakeholders responded to the registry with excitement and enthusiasm, and entities – from governments and schools to banks and oil majors – registered commitments that simultaneously advanced their interests and the cause of sustainable development generally. This response was encouraging and certainly suggested a high level of support and embrace from stakeholders. Commenting on the potential impact of the Registry, the NRDC (2012) observed:

> Rio+20 could be the first summit where a web-based platform will be the major output. By making promises in Rio to specific actions, presidents, prime ministers, CEOs and other leaders can demonstrate a new determination to speed the transition to a sustainable economy and to be held accountable. The creation of the registry alone would be an important and encouraging signal that our leaders are willing to embrace new technologies and approaches to tackle the huge and growing challenges we face.

Not all the commentary was positive; Richard Black, a senior BBC environmental journalist, whose coverage of Rio+20 had been much more balanced than that of most of the other journalists, in a very critical article on the Rio+20 voluntary commitments dismissed it all with this question: "Before Rio+20, just about every environmental indicator was pointing in the wrong direction . . . So in all the voluntary commitments made at Rio, precisely what is there that will turn any of these trends around?" (Black, 2012). Several of the Major Groups were also concerned that these commitments do not substitute for member states' commitments.

Black's question, although legitimate in any serious inquiry, may have failed to take into account the Sustainable Energy for All initiative; the Higher Education Sustainability Initiative; the policy and other commitments made by the SIDS; the concrete actions, pledges and programmes to be undertaken by stakeholders; the new finance committed to the development process; and the extent to which these could together contribute across all sectors and at a global level to turn around negative trends. Of greatest concern, however, is the fact that the commentator ignored the matrix for measurement of the level of commitments versus the level of attainment and the peer review mechanism for ensuring accountability. Only time will tell whether he was right.

## Box 4.2 A sample of the commitments

- The Bank of America pledged US$50 million over a ten-year period to initiatives and business activities relating to the Sustainable Energy for All initiative and its objectives of increasing renewable energy, energy access and energy efficiency.
- The Bank of America pledged also US$100 million in grants and programmes related to investments to non-profit organizations, community development organizations, financial institutions and other NGOs promoting low-carbon resources, conservation solutions, and reduction of energy consumption water and waste.
- BPW International, the International Federation of Business and Professional Women, launched a Sustainability Business Award.
- Chevron pledged US$50 million to improve living standards in the Niger Delta.
- Cx Catalysts pledged to put initiatives in place to empower female entrepreneurs: 100 women in Caribbean SIDS, 500 in Guatemala, 100 in Turkey, 100 in the Pacific, 50 in Spain and 5,000 in Africa.
- Darden Restaurants (Publix Supermarkets) made pledges on fisheries preservation and rebuilding, and on catalysing industry to rebuild world fisheries stocks.
- Earth Charter International pledged an initiative on corporate sustainability reporting.
- Rent-a-Car pledged to plant 50 million trees and use more electrical cars in its fleet.
- The energy company Masdar (in the United Arab Emirates) pledged to build 600 solar houses.
- The mining and metals company Rio Tinto pledged greater workforce diversity, including an increasing number of women in the workforce and management structure.
- Three hundred universities and business schools pledged to sign the declaration committing to implementing the Higher Education Sustainability Initiative's principles in their institutions (Euromed Management, 2012b).
- The Walt Disney Company pledged that it would invest in ways of shrinking ecological impacts by implementing initiatives in energy, carbon emissions, ecosystems, waste and water (UN DESA, 2012b).

The critical task now is to ensure that the voluntary commitments are fully implemented. The standards for measurement of actions and execution which UN DESA insisted should be included in all pledges has the potential to hold pledgers accountable and make the whole process more open to public scrutiny. The inclusion of a transparent mechanism for accountability was a strong and valuable addition to the multilateral process; broadening access and oversight by stakeholders and creating a new space in which all stakeholders could interact on development activity. This issue of the post-conference policing and follow-up is pivotal to the success of the Commitments. In this regard, Céline Ramstein (2012), a young diplomat invited by the French government to be part of France's delegation to the Rio+20 process, has proposed the establishment of a multi-stakeholder advisory board to support the implementation of the voluntary commitments in a transparent manner. This valuable suggestion can contribute significantly to the success of the commitments. As an important follow-up, UN DESA has created a form on the Sustainable Development Knowledge Platform that allows partners to report on progress in following through their voluntary commitments, which will increase transparency.

## National Councils for Sustainable Development and economic and social committees

Agenda 21, as agreed at the 1992 Rio Earth Summit, asked all countries to establish National Sustainable Development Strategies (NSDS) and to develop the necessary national governance structures to put them in place. At Rio+20 there were meetings of the national economic and social committees from around the world at which they came together to commit to making sustainable development a central, overarching theme of their work. The various national councils for sustainable development also came together in 2012 to make a similar commitment and to launch a new network intended to share knowledge and information about their activities and to identify areas for collaborative action. This new network can have a significant role in sharing knowledge and capacity, and possibly even in offering a linkage between whatever new global body is created for sustainable development and the activities at sub-national and local government level. The network is managed by Stakeholder Forum.

> The Global Network of National Councils for Sustainable Development (NCSDs) aims to strengthen and enhance the work of the highest level sustainable development bodies at the national level. To achieve this, the Global Network provides information on the form and function of the numerous NCSDs across the world, as well as providing a space for these Councils to exchange experiences, ideas and discuss ways to cooperate.
>
> (UNCCD, n.d.)

## Sub-national and local government

Regions and cities play a vital role in the socio-economic development of all countries. Inclusive, efficient, productive and sustainable regions often provide the backbone for national economic growth and for the well-being of the population. Cities have huge ecological footprints to support their demand for food, water, fibres, timber, energy, landfill and space. Urban infrastructure requires biologically productive land across the world. As UN Secretary-General Ban Ki-moon declared in the run-up to Rio+20, the battle to create a sustainable path for all will be won or lost in the cities. Cities also consume more than 75 per cent of the world's natural resources, use approximately 75 per cent of the world's energy and are responsible for 75 per cent of carbon emissions. By 2050, 70 per cent of the world's 9.6 billion people are expected to live in cities, compounding these effects and further growing the numbers living in slums.

> In countries of the Organisation for Economic Co-operation and Development (OECD), urban areas produce 80 percent of the Gross Domestic Product (GDP). It is estimated that between 2005 and 2025 some 200 trillion dollars will be spent globally on fixed urban assets. If we are to transition to a global Green Economy, it is crucial that these investments are made in line with an inclusive green economy. The territorial dimension of a Green Economy capable of unleashing the potential of a given territory, of creating local skills and of generating local and subnational resilience should also be highlighted.
>
> (Local Authorities Major Group, 2012)

Since 1992, local and sub-national governments all over the world have been fully involved in sustainable development processes. Their involvement was inspired by chapter 28 of Agenda 21, which in particular said:

> Local authorities construct, operate and maintain economic, social and environmental infrastructure, oversee planning processes, establish local environmental policies and regulations, and assist in implementing national and subnational environmental policies. As the level of governance closest to the people, they play a vital role in educating, mobilizing and responding to the public to promote sustainable development.
>
> (United Nations, 1992)

In 1990 a global network for local government was set up, with initially over 200 local authorities. It was called the International Council for Local Environmental Initiatives, and under its first Secretary-General, Jeb Brugmann, it played a significant role in the development of Agenda 21 and the promotion of the local implementation of sustainable development, as stated in chapter 28 of Agenda 21. With 20 years of experience with local Agenda 21, this organization, now renamed ICLEI Local Governments for Sustainability, produced a local authorities' "Self-Assessment of Local Agenda 21 Toolkit" that looked at areas such as territorial sustainable development strategies, public–private partnerships for green innovation, social action plans, green public procurement initiatives, and decentralized cooperation projects by these levels of government.

All these initiatives have evolved significantly over the past years, and several of them are carried out on the ground in priority areas and emerging challenges highlighted in Rio+20's "The Future We Want." These initiatives have delivered good results, are replicable and can be scaled up. This demonstrates that the efforts and actions of a country can become stronger when transposed and complemented by governmental stakeholders, such as federated states, regions, cities and other local authorities (Local Authorities Major Group, 2012). Subnational and local governments can help deliver sustainable cities and human settlements, which can boost a country's social, economic and environmental performance.

The three co-organizing partners of the Local Authorities Major Group for Rio+20 (LAMG) were ICLEI Local Governments for Sustainability, the Network of Regional Governments for Sustainable Development (nrg4SD) and United Cities and Local Governments (UCLG). They accredited over 1,000 participants to the Rio+20 Conference and organized a Global Town Hall, the World Summit of States and Regions and the Urban Summit – where subnational and local governments renewed their commitments to implementing sustainable development in their respective territories and formed new partnerships to carry this work forward with other levels of government, the private sector and academia.

## Future Earth

A new ten-year international research initiative on global sustainability, Future Earth, was launched at Rio+20. The programme aims to develop "the knowledge required for societies worldwide to face challenges posed by global environmental change and to identify opportunities for a transition to global sustainability."

A unique Science and Technology Alliance for Global Sustainability was formed to establish Future Earth. This alliance includes global science players: the International Council for Science (ICSU), the International Social Science Council (ISSC), the Belmont Forum, the United Nations Educational, Scientific and Cultural Organization (UNESCO), the United Nations University (UNU), UNEP and the World Meteorological Organization (WMO) (the last of these as an observer). Future Earth aims to mobilize thousands of scientists around a research programme on global sustainability. But it also wants to create an international platform between science, policy and other stakeholders. It will play a key role in coordinating the scientific community in relation to international processes such as the one for elaborating the Sustainable Development Goals, the Intergovernmental Panel on Climate Change and the Intergovernmental Platform on Biodiversity and Ecosystem Services. As Diane Liverman, one of the co-chairs of a transition team for Future Earth, said, "We need a new approach to address the critical challenges of global environmental change and sustainable development which is more interdisciplinary, more international, more collaborative and more responsive to the users of research" (quoted in ICSU, 2012).

Future Earth aims to deliver:

- solution-oriented research for sustainability, linking environmental change and development challenges to satisfy human needs for food, water, energy and health;
- science input to the development and implementation of Sustainable Development Goals;
- effective interdisciplinary collaboration across natural and social sciences, humanities, economics, and technology development to find the best scientific solutions to multifaceted problems;
- timely information for policy-makers by generating knowledge to support existing and new global and regional integrated assessments;
- participation of policy-makers, funders, academics, business and industry, and other sectors of civil society in co-designing and co-producing research agendas and knowledge;
- increased capacity-building in science, technology and innovation, especially in developing countries and engagement of a new generation of scientists.

It will focus on three main themes to develop the integrated research for global sustainability:

- dynamic planet: observing, understanding, projecting Earth and societal system trends, drivers and processes, and their interactions; anticipating global thresholds;
- global development: providing the knowledge on the pressing challenges to provide sustainable, secure and fair stewardship of food, water, health, energy, materials and other ecosystem services;
- transforming towards sustainability: understanding transformation processes and options; assessing how these relate to human values, emerging technologies and social and economic development pathways; and evaluating strategies for governing and managing the global environment across sectors and scales.

Future Earth offers a real chance to help policy-makers understand the challenges ahead and what their options for overcoming those challenges might be. It will also help science to underpin the new development goals to be agreed for beyond 2015.

## Education and training

Education is an important vehicle for mainstreaming sustainability. Rio+20 saw the launch of the Higher Education Sustainability Initiative (HESI), one key voluntary initiative developed in the office of the Executive Coordinators for Rio+20. It is targeted at institutions of higher learning and in its initial conception was aimed at universities and business schools. The crafting of the HESI came about in an interesting way. In speeches during 2011, Elizabeth Thompson, one of the United Nations' two executive coordinators for Rio+20, on three different occasions when addressing the business sector, business schools and educational institutions pointed out the need for an educational initiative (1) to advance sustainable development, (2) to increase outreach to the business sector, and (3) to get business schools and universities both to teach and to practise sustainable development. Dr Mohan Peck, then a senior sustainable development officer within UN DESA, was assigned to work with Thompson. Peck and Thompson framed possible core principles for such an initiative. Peck then liaised with a number of UN agencies that liked the idea, and worked with Thompson to expand on the core principles.

In order to achieve support and signatories from within the academic community, the initiative received a strong push from a number of UN agencies. including the UN Academic Initiative, the UNDP, UNEP, UNESCO, UNU and Global Compact-Principles for Responsible Management Education (Global Compact-PRME, pronounced "prime"). Jonas Haertle, the lead officer of PRME (United Nations Visitors Centre, 2012), worked with their membership in creating interest in HESI, building understanding and support for the initiative, and hosting the launch at Rio+20. These collaborating UN agencies framed an HESI Declaration to which universities would sign up and by which they would agree to be bound to the initiative. Jean-Christophe Carteron (Bentley University, 2012) of Euromed Management worked closely and enthusiastically with Peck in the drafting of the Declaration and was instrumental in encouraging European institutions to join the Initiative.

The HESI was first made public on March 1, 2012. In a letter disseminated to tertiary institutions, Thompson indicated:

> The HESI[,] which is targeted at universities and business schools, is aimed at creating an enlarged understanding of sustainable development and its implications for how we live[,] work and do business. It is hoped that the initiative will result in larger numbers of graduates globally who understand sustainable development, its meaning for their discipline and area of enterprise, natural resource consumption and the methods through which market and shareholder value might be [created or] enhanced by incor-

> porating sustainable practices as central to how they operate. The core message is that those who occupy the planet are all shareholders of Earth Incorporated, that we are all potential practitioners of sustainability and that by investing in natural resource capital, the return on investment can be high for the planet and the citizens, countries, and companies which occupy it.
>
> (Thompson, 2012)

It is intended to increase the numbers of practitioners of sustainability. In synopsis, the HESI invites institutions of tertiary education to commit to the following (Euromed Management, 2012b):

- Reducing their large ecological footprints by practising the 3Rs (reduce, recycle, reuse) and proper waste management, and instituting energy and water use policies.
- Greening their campuses including their buildings, retrofitting where necessary.
- As major influences on local economies wherever they are located, greening their supply chains and procurement policies.
- Supporting research into green technologies.
- Producing and promoting literature that compiles best practices in sustainability and creating a body of work around sustainability issues, including case studies that relate to business practice.
- Teaching sustainable development as a core module across all disciplines in the university so that every graduate has an understanding of issues in the development dialogue, such as the concept of planetary boundaries, climate change, sustainability, natural resource capital and protection as it relates to his or her area of learning or enterprise. This will produce not only better-rounded graduates but also practitioners of sustainability in all disciplines.
- Supporting sustainable methods of transport in and around campus.
- Disseminating information and sustainable development material among their international networks and local communities.

At the time the work was being done to develop the HESI and make it ready for the public, those involved in the effort did not know that they had created what would be the single most successful initiative at Rio+20, the one that would attract the largest number of voluntary commitments. Just under 300 universities and business schools from more than 50 countries (Euromed Management, 2012b) in every part of the globe committed to the initiative by signing the Declaration. This represented an excellent effort by the promoters of an initiative that had been developed and made public a mere ten weeks before the Rio+20 Conference. The HESI was endorsed by a further 43 organizations (Euromed Management, 2012c). The institutions of higher learning were also provided with a list of resources on which they could rely for support in seeking to fulfil the HESI's mandate.

Besides the innovative and successful HESI, education was also covered in the Rio+20 Outcome Document in paragraphs 234 and 235, which relate to HESI:

> We strongly encourage educational institutions to consider adopting good practices in sustainability management on their campuses and in their communities with the active participation of, inter alia, students, teachers and local partners, and teaching sustainable development as an integrated component across disciplines.
>
> We underscore the importance of supporting educational institutions, especially higher educational institutions in developing countries, to carry out research and innovation for sustainable development, including in the field of education, to develop quality and innovative programmes, including entrepreneurship and business skills training, professional, technical and vocational training and lifelong learning, geared to bridging skills gaps for advancing national sustainable development objectives.
>
> (UN DESA, 2012c)

The question that might now be asked is: now that the HESI has got past its first year, apart from attracting more signatories what should be the next steps in its development?

It should be beneficial for universities participating in the HESI to use it as a marketing tool to help enhance the university's brand and attract students. This would require the development of some objective standards to encapsulate the qualifying criteria, a matrix with set benchmarks and targets that would allow for comparative analyses, and objective measurements to see what has been achieved. The second push required by the UN system engaged on educational matters, and specifically the HESI, is to try to get the practice, promotion and teaching of sustainability included among the criteria used by agencies that rank business schools and universities. This is not a far-fetched notion. As the green economy, natural capital accounting and related considerations find themselves into business plans, social, financial and economic instruments, and as a Sustainable Development Index is used, as opposed to the purely economic considerations of GDP, there will be an increased need for graduates who are knowledgeable and skilled in sustainability. It is also necessary to view the teaching of sustainability at the tertiary level only as a starting point in the spread of the message and benefits of sustainability, which should be taught at every level of the education system so as to give impetus to the mainstreaming of sustainability in every stream of society and economy.

A successful HESI will mean a new body of literature and case studies from which all stakeholders can draw, an increase in the use of innovation and green technologies, in sustainable consumption and production, and in the development of a global green economy with business and education at its core. It could also result in the teaching of short modules, especially in an online format, to all persons interested in sustainability, from housewives to business-people. UN DESA (2012a) explains the HESI on its website in this way:

> Education for sustainable development aims at enabling everyone to acquire the values, competencies, skills and knowledge necessary to contribute to building a more sustainable society. This implies revising teaching content to respond to global and local challenges. It should also promote teaching methods that enable students to acquire skills such as interdisciplinary thinking, integrated planning, understanding complexity, cooperating with others in decision-making processes, and participating in local, national and global processes towards sustainable development.

Education has long been a tool of social and economic transformation and upward social mobility for individuals. We have increasingly come to live in what Peter Drucker has called the "knowledge economy," in which education and specialist skills characterize workers and are important drivers of competitive advantage, business opportunity, innovation and social and economic growth.

## The World Congress on Justice, Governance and Law for Environmental Sustainability

One of the events that brought together a new stakeholder group was the World Congress on Justice, Governance and Law for Environmental Sustainability, which gathered over 250 chief justices, attorneys general and auditors general. Organized by Bakary Kante, director of the Division of Environmental Law and Conventions, this successful initiative was the first ever meeting of chief justices of the world's supreme courts and produced a "Rio+20 Declaration on Justice, Governance and Law for Environmental Sustainability." The Declaration contained the following key recommendations (UNEP, 2012b):

- Without adherence to the rule of law and open, just and dependable legal orders the outcomes of Rio+20 will remain unimplemented.
- An independent Judiciary and judicial process is vital for the implementation, development and enforcement of environmental law, and members of the Judiciary, as well as those contributing to the judicial process at the national, regional and global levels, are crucial partners for promoting compliance with, and the implementation and enforcement of, international and national environmental law.
- Environmental law is essential for the protection of natural resources and ecosystems and reflects our best hope for the future of our planet.
- Environmental litigation often transcends national jurisdictions. We need more effective national and international dispute settlement systems for resolving conflicts.

- Environmental sustainability cannot be achieved without good quality data, monitoring, auditing and accounting for performance.
- Environmental and sustainability auditing ensures transparency, access to information, accountability, and efficient use of public finances, while protecting the environment for future generations.
- Judges, public prosecutors and auditors have the responsibility to emphasize the necessity of law to achieve sustainable development and can help make institutions effective.
- Scientific information and knowledge is a central foundation of effective compliance with and enforcement of environmental obligations.
- States should cooperate to build and support the capacity of courts and tribunals as well as prosecutors, auditors and other related stakeholders at national, sub-regional and regional levels to implement environmental law, and to facilitate exchanges of best practices in order to achieve environmental sustainability by encouraging relevant institutions, such as judicial institutes, to provide continuing education.
- The existing international governance institutions to protect the global environment should be strengthened. We must create modern institutional structures capable of building networks and improved sharing of decision-making. There is an urgent need to give consideration to transforming UNEP to effectively lead and advance the global policy and law-making agenda for the environment within the framework of sustainable development.

## Initiatives on transparency and access to information

Rio+20 also endorsed Principle 10 of the Rio Declaration, which promotes public participation in decision-making and access to justice and information. This is perhaps best expressed in law through the United Nations Economic Commission for Europe's "Convention on Access to Information, Public Participation in Decision-making and Access to Justice in Environmental Matters," also known as the Aarhus Convention. This was a very strong recommendation and in addition had strong support among many other stakeholders.

On July 25, 2012 the government of Chile, supported by Costa Rica, the Dominican Republic, Ecuador, Jamaica, Mexico, Panama, Paraguay, Peru and Uruguay, tabled a "Declaration on the Application of Principle 10" in the United Nations, based on their verbal statement on June 27. The Declaration committed the countries to seek a regional convention on Principle 10:

> Commitments must be made to ensure the full exercise of rights of access to information, participation and justice regarding environmental issues as enshrined in Principle 10 of the Rio Declaration of 1992. To this end, we declare our willingness to launch a process to explore the feasibility of adopting a regional instrument, ranging from guidelines, workshops and

> best practices to a regional convention open to all countries in the region and with the meaningful participation of all concerned citizens.
>
> (Chile, 2012)

It was intended that there would be an ongoing network under the leadership of UNEP that would focus on providing information, the development, and implementation of environmental law and capacity-building.

## Sustainable Energy for All (SE4All) and commitments and achievements on energy at Rio+20

The Sustainable Energy for All (SE4All) initiative was one of the most successful outcomes of Rio+20. That success is evidenced in two ways: in the Outcome Document and in the voluntary commitments. In paragraphs 125–129 of the Outcome Document, member states commit to the three objectives of the SE4All initiative (increasing renewable energy, energy efficiency and energy access) and to advancing energy as a key development tool. At Rio+20 the initiative attracted significant funding, a large number of countries committed to it and several commitments from private-sector entities were also recorded.

A press release issued by the United Nations on June 21, 2012 (United Nations, 2012d) during the high-level segment of the conference gave greater detail on what the initiative had accomplished since its launch and warrants repetition nearly in its entirety. The commitments listed below are indicative of the list but are not comprehensive:

- More than 50 governments from Africa, Asia, Latin America and the Small Islands Developing States have engaged with the initiative and are developing energy plans and programmes.
- Businesses and investors have committed over US$50 billion to achieve the initiative's three objectives. Tens of billions of dollars have been committed by other key stakeholders – governments, multilateral development banks, and international and civil society organizations – to catalyse action in support of the initiative.
- More than a billion people will benefit from Sustainable Energy for All's public- and private-sector commitments. Most beneficiaries in developing countries will gain improved access to energy through grid extension and off-grid solutions, as well as scaled-up renewable energy sources, increased investment and improved energy policies.

Additionally, the government of Brazil demonstrated outstanding host-country leadership by committing to invest US$4.3 billion to achieve universal energy access in Brazil by 2014. As a global leader in sustainable energy, Brazil will invest an additional US$235 billion over ten years in renewable energy, mainly in hydropower and biofuels but also in biomass and wind power, and will produce total energy savings of 9 per cent in 2030 from the electrical and transportation sectors.

Below is a partial list of commitments to action made to date in support of Sustainable Energy for All:

- US$700 million has been committed for a Partnership Initiative on Energy.
- US$80 million has been committed by the United States to the replenishment of the Global Environment Facility (GEF).
- US$970 million has been committed over three years in sanitation and water projects.
- US$43 million is to be invested in energy in 2013.
- The European Commission of the European Union announced a commitment of US$521 million to support sustainable energy.
- Norway has pledged US$140 million over five years to scale up access to sustainable energy in Ethiopia, Kenya and Liberia.
- The United States has announced a partnership with African nations, with US$20 million in funding for clean energy projects in Africa.
- Eight multilateral development banks have pledged to invest US$175 billion over the next ten years to support the creation of sustainable transport systems.
- Private-sector companies have pledged to contribute US$50 billion to a plan to provide energy to the entire global population by 2030.

### *Governments and country actions*

There were in addition a number of government and country actions announced at Rio+20. These included the following:

- Ghana, one of the first countries to partner with the SE4All initiative, has developed a national energy action plan to support capacity development and innovative financing mechanisms.
- Countries initiating or completing similar assessments include Bangladesh, Kenya, Mozambique, Nepal, Tajikistan, Uruguay and Vietnam.
- Twenty SIDS have agreed to work towards universal access to energy, switching to renewable energy and reducing dependence on fossil fuels. For example, the government of Barbados will increase its use of renewable energy to 29 per cent of all electricity consumption by 2029.
- The European Union announced the initiative "Energizing Development," which will provide access to sustainable energy services to 500 million people by 2030 and create a technical assistance facility supported by approximately US$63 million over the next two years.
- Norway announced a new commitment of approximately US$140 million over five years to scale up access to sustainable energy in Ethiopia's rural areas, to replace kerosene lamps with solar alternatives in Kenya and to

support Liberia's development of a strategic energy and climate plan, among other ongoing initiatives.

- The United States announced its strong support for the SE4All Global Action Agenda and plans to provide US$2 billion in grants, loans and loan guarantees for policy and regulatory development, public–private energy technology partnerships, and loans and guarantees to leverage private investment in clean energy technology.

### *A new engagement for business*

Rio+20 had the largest ever number of private-sector participants for a UN conference, with over 1,600 registered participants. A global transition to a new green economy will require substantial input and cooperation from the private sector, albeit with the recognition that that transition will be different depending on national circumstances. Cognizant of this reality, the Global Compact noted the following on its website:

> There has never been greater urgency to secure a better environmental, economic and social future for all.
>
> The private sector has a critical role to play in this endeavour, and is increasingly putting sustainability – defined as a company's delivery of long-term value in financial, social, environmental and ethical terms – on its agenda.
>
> (United Nations Global Compact, 2012)

At the Rio+20 Conference the Global Compact sponsored a Corporate Sustainability Forum during which 60 sessions of meetings for the private sector over a four-day period took place. Three thousand businesspeople attended the forum and engaged on the following topics (IISD, 2012):

- energy and climate;
- water and ecosystems;
- agriculture and food;
- social development;
- urbanization and cities;
- economics;
- finance for sustainable development;
- "hurdles to achieving scale in sectors including, agriculture, cement, chemicals, consumer goods, energy/power, forestry, materials, oil and gas, and transport" (IISD, 2012).

These meetings intended to address issues such as "the right incentives and enabling environment," in addition to showcasing "innovative public–private partnerships, business contributions and new commitments" and to developing "recommendations for government action." The Corporate Sustainability Forum became one of the positive outcomes of Rio+20 and precipitated significant interest and dialogue among businesspeople (Steven and Evans, 2012). A specific platform for business emerged from the forum. Added to the Global Compact's work and that of other organizations at Rio+20, some principal components of a platform for business have emerged:

- endorsement of a Green Industry Platform by 70 business leaders;
- publication of a communiqué by 45 CEOs supporting global water sustainability;
- agreement by 16 companies to develop principles of good practice and policy on sustainable agriculture;
- launch of a Social Enterprise Investment Framework for business;
- agreement by some 200 Brazilian CEOs to work towards a green and inclusive economy;
- a new platform to advance transparency and the inclusion of sensitive data in business practice supported by 400 CEOs.

The private sector was also asked to announce commitments; some were as follows:

- Microsoft committed to going carbon neutral.
- The Global LPG Partnership committed to scaling up access to liquefied petroleum gas (LPG) for cooking by to 50 million people by 2018.
- GDF Suez committed to investing in approximately 50 local energy entrepreneurship projects in developing countries by 2020.
- Statoil committed to continuing to implement its "no production flaring" policy and to work with the World Bank to expand the Global Gas Flaring Reduction partnership.
- Eni earmarked approximately US$5 billion to achieve its gas-flaring and carbon intensity reduction goals and committed to spending approximately US$32 million on low-voltage power lines in the Republic of the Congo, benefiting 140,000 people and their communities.
- The Renault-Nissan Alliance committed approximately US$5 billion to commercialize affordable zero-emission vehicles, adding five distinct models by the end of 2012. It plans to have 1.5 million zero-emission vehicles on roads through the fiscal year 2016.
- Light Design committed to providing solar lamps to 30 million people in more than 40 countries by 2015.
- Infosys, a provider of business and outsourcing services in 32 countries, committed to reducing energy consumption by 50 per cent and sourcing 100 per cent of electricity from renewables by 2018.

- FEMSA, a consumer company in Latin America, committed to cover 85 per cent of the energy needs of its Mexican operations through renewable energy sources, such as wind and solar power, by 2013.
- The Global Sustainable Electricity Partnership pledged to install 50,000 solar lanterns that will provide clean electricity to off-grid households and electrification projects, including Electrification Roadmaps for Southern Africa, which aims to provide access for 500 million people by 2025.
- Masdar, the energy company of the United Arab Emirates, committed to setting up renewable energy projects in Tonga and Afghanistan and created the Global High School Prize, awarding up to US$100,000 to five schools that submit winning cases for how to improve their school's energy footprint.

### *Development banks and institutions*

A number of international financial institutions also made significant commitments during the Rio+20 Conference. These included the following:

- Bank of America pledged to set a ten-year US$50 billion environmental business goal, estimating that this should equate to approximately US$35 billion in SE4All focus: energy, energy efficiency and energy access.
- Multilateral development banks committed more than US$30 billion towards achieving SE4All's three objectives. The African Development Bank pledged US$20 billion in energy by 2030, and will draw an additional US$80 billion by partnering with private and public enterprises for regional projects that benefit multiple countries and small and medium-scale energy projects that increase access to electricity in rural regions, including through its Sustainable Energy Fund for Africa.
- The European Bank for Reconstruction and Development committed US$8 billion in energy efficiency projects in Eastern Europe and Central Asia over the next three years.
- The Organization of the Petroleum Exporting Countries (OPEC) Fund for International Development (OFID) committed a minimum of US$1 billion to finance its "Energy for the Poor" initiative and its work against energy poverty.
- The ClimateWorks Foundation committed approximately US$15 million in supporting and matching funds over three years to assist Brazil, China, India and Mexico to identify, adapt, and implement proven best-practice energy efficiency and renewable energy policies.

### *Actions and commitments by the NGO community, artists, academia and individuals*

Rio+20 also saw key stakeholders making substantial commitments. These included:

- Members of the rock band Linkin Park committed to a "Power the World" campaign.
- The UN Foundation, which is coordinating the Energy Access Practitioner Network, committed to bringing together more than 500 individuals from 350 organizations.
- The Renewable and Appropriate Energy Laboratory at the University of California, Berkeley, pledged to share tools for modelling low-carbon sustainable energy futures and to chair a network of academic partner institutions committed to building educational curricula around achieving the initiative's objectives.
- The Energy and Resources Institute (TERI) in India committed to expanding lighting services to households in several developing countries, using solar and other clean energy technologies, impacting an estimated 10 million lives by 2018.
- The Institute of Electrical and Electronics Engineers (IEEE) in the United States and 40 other professional associations have pledged to mobilize their 2 million members worldwide in support of the SE4All initiative.

### *Registry of Voluntary Commitments*

The Compendium/Registry of Voluntary Commitments for Rio+20 included numerous commitments on areas such as transport, energy, water, cities, technology and innovation (UN DESA, 2012b). These all provide points of entry for the involvement of and investment by the business sector. The financial commitments made by governments, and towards the SE4All initiative particularly, will require an active and sustained role by business. What the Global Compact hoped to engender was a level of dialogue and commitment between business, governments and other stakeholders, as well as exciting the interest of the private sector in the benefits of engaging in a multilateral agenda at the heart of which is a green economy.

### *Business investment and partnership*

Given the current global financial challenges, governments need new investment and capital via partnerships with private-sector entities such as through Build Operate Lease Transfer (BOLT) and Build Own Operate Transfer (BOOT) mechanisms. The private sector will also have to lead on green technologies and innovation, green investment, and the creation of green jobs and decent work. Governments' creation of appropriate incentives, and of enabling a legal, regulatory, policy and fiscal environment to facilitate the business sector's role and investment in the green economy, was an issue on which those in the private sector needed to reflect and discuss among themselves and with other stakeholders.

The business sector conceded the challenges of an era of natural resource shrinkages and rising prices; the need to associate brands and brand profiles with sustainability; the need to enhance corporate social responsibility practice; and the need to increase sustainability reporting as well as shareholder value and competitive advantage using sustainability as the base. So too does the natural capital accounting effort being pursued by the World Bank, which was supported by over 50 countries and 86 private companies (World Bank, 2012a, b). The International Chamber of Commerce also developed a green growth road map and initiatives in relation to the green economy (International Chamber of Commerce, 2012).

### *The Natural Capital Declaration*

Another output from Rio+20 was the Natural Capital Declaration, which is a declaration by 37 CEOs of institutions in the wider financial sector (banks and insurance companies). It describes itself as "a global statement that demonstrates the commitment of the financial sector to work toward integrating natural capital criteria into financial services and product" (UNEP, 2012a).

The criteria it establishes include:

- the implementation of natural capital impact disclosure requirements for companies;
- enforceable fiscal measures conducive to the valuation of natural capital;
- the implementation of the Convention on Biological Diversity;
- the setting of an example by governments through the requirement for public spending, and procurements to report and eventually account for the use of natural capital.

Part of the text reads as follows (UNEP, 2012a):

> Anticipating that such a framework will emerge, and noting that no methodology yet exists to adequately report or account for natural capital in the global financial system, we the endorsing financial institutions wish to demonstrate leadership by undertaking to collaborate globally through working groups and engagement with our customers, investee companies, suppliers, civil society, and other stakeholders as appropriate to:
>
> 1. Build an understanding of the impacts and dependencies of natural capital relevant to our operations, risk profiles, customer portfolios, supply chains and business opportunities;

2. Support the development of methodologies that can integrate natural capital considerations into the decision-making process of all financial products and services – including in loans, investments and insurance policies. We recognise that given the diversity of the financial sector, embedding natural capital considerations will differ across asset classes and types of financial institutions. We therefore aim to build on work undertaken through other initiatives, such as the UN-backed Principles for Responsible Investment, the Equator Principles, the United Nations Environment Programme Finance Initiative (UNEP FI) Principles for Sustainable Insurance, and The Economics of Ecosystems and Biodiversity (TEEB), so that we can develop methodologies to:

   A: Apply a holistic approach to evaluating bonds and equities through the integration of natural capital considerations in environmental, social and governance (ESG) risk analysis in short, medium and longterm growth forecasts of investee companies;
   B: Systematically consider and value natural capital in the credit policies of specific sectors, including commodities, that may have a major impact on natural capital either directly or through the supply chain;
   C: Systematically consider and value natural capital in core insurance business strategies and operations including risk management, risk underwriting, product and service development, claims management, sales and marketing, and investment management.

3. Work towards building a global consensus for the integration of natural capital into private sector accounting and decisionmaking; supporting, when appropriate, the related work of the TEEB for Business Coalition, and other stakeholders.
4. Collaborate, when appropriate, with the International Integrated Reporting Committee and other stakeholders to build a global consensus around the development of Integrated Reporting, which includes natural capital as part of the wider definition of resources and relationships key to an organization's success.

### *Corporate responsibility and greater business accountability*

During the preparation for Rio+20 there was a push for the introduction of mandatory sustainability reporting for companies listed on stock exchanges. The proposal was that companies would report or explain, a concept already in place in stock exchanges in Brazil, Malaysia and South Africa, and was in the process of being adopted in the European Union through changes in the Accounting Directive. The proposal was supported by a wide range of stakeholders, including large corporations, such as Dow, and financial institutions, such as

Aviva, HSBC, the Corporate Sustainability Reporting Coalition (CSRC) and NGOs. Despite this, paragraph 47 became diluted.

Post Rio+20, an example of the desire of business to become immersed in the green economy and sustainability agenda may be seen in a letter written by the CSRC, convened by Aviva Investors, to the UK Prime Minister, David Cameron, who, along with the Presidents of Liberia and Indonesia, co-chaired the UN Secretary-General's High-Level Panel on the Post-2015 Development Agenda. Their letter provided the panel members with a set of points guide to help their deliberations. Their primary points (CSRC, 2013) were that:

- The private sector could have a critical and transformative role in implementing a post-2015 development framework linking environmental sustainability and poverty eradication; hence corporate sustainability reporting should be seen as central to the development policy process and programmes.
- Corporate sustainability reporting exemplified in disclosure and transparency policies will enhance both corporate and government performance; cooperation between government, business and society will in turn lead to a stronger, more inclusive post-2015 development framework and agenda.
- By accumulating, assessing, measuring and disclosing governance, social and environmental information, companies would be best able to manage their impacts.

Rio+20 provided a pivotal shift in the engagement and relationship of stakeholders in business, government and society as it relates to the multilateral development agenda. The statements, efforts and initiatives by the business sector, as well as business's positive engagement in most cases with other stakeholders at Rio+20 and since then, augur well for the transition to a global green economy and for business embracing its role in social development, poverty eradication and their myriad interrelated components. Business's commitment at and following Rio+20, and more recently the CSRC's letter to the High-Level Panel on the Post-2015 Development Agenda, are powerful indicators of the new engagement as a willing development partner that some businesses are showing.

## Financial commitments

There were no financial commitments made at the Earth Summit of 1992. As well as policy discussion, Rio+20 saw governments and intergovernmental organizations make a number of financial commitments to help fund the implementation of "The Future We Want."

Most donors had come to Rio+20 not willing to contribute substantial new funds for implementation. No new country had reached the UN target of 0.7 per cent GDP for overseas aid in the past ten years. The club of Denmark,

Luxembourg, Netherlands, Norway and Sweden may, however, have a new member in 2013, since the United Kingdom is well on track. The Netherlands, however, fell below 0.7 per cent post Rio+20. Only two days after the Rio+20 Conference, the UK overseas development minister, Andrew Mitchell, reaffirmed the UK pledge: "The key thing is we are standing by our commitments and from 2013 we will reach 0.7 per cent – that's the absolute commitment" (Mitchell, 2012).

Other government financial commitments made at the Rio+20 conference included the following:

- Australia committed to doubling its support to the Coral Triangle Initiative, with US$8.14 million plus an additional US$25.4 million in funding for fisheries and climate change adaptation on the Pacific.
- Brazil committed US$6 million to UNEP and US$10 million towards climate change challenges in Africa, LDCs and SIDS.
- China committed US$6 million to UNEP.
- The European Union's European Commission announced US$521 million to support sustainable energy.
- Japan announced funding for a three-year programme of disaster risk reduction (no figure available yet).
- Norway announced US$140 million over five years to scale up access to sustainable energy in Ethiopia, Kenya and Liberia.
- The United Kingdom announced US$241 million for the International Fund for Agricultural Development's Adaptation for Smallholder Agriculture Programme.
- The United States announced a partnership with African nations, with US$20 million in funding, for clean energy projects in Africa.
- Eight multilateral development banks pledged to invest US$175 billion over the next ten years to support the creation of sustainable transport systems.

The situation ten years previously at the World Summit on Sustainable Development in Johannesburg, a summit held in the long shadow of 9/11, makes an interesting comparison. The commitments made at Rio+20 in 2012 are much larger than those made at Johannesburg at 2002. The commitments at the WSSD were as follows (Dodds and Nayer, 2012):

- The Asian Development Bank committed US$5 million to UN Habitat and US$500 million in fast-track credit for the Water for Asian Cities Programme.
- The European Union committed US$700 million to a Partnership Initiative on Energy and US$80 million to the replenishment of the GEF.
- The United States committed US$970 million over the following three years in sanitation and water projects, US$43 million to be invested in energy in 2003, US$2.3 billion through 2003 on health, US$90 million in

2003 for sustainable agriculture programmes, and US$53 million for forests between 2002 and 2005.

### *A strategy for mobilizing financing for sustainable development*

There had been talk in the run-up to Rio+20 of the possibility of the eurozone introducing a tax on financial transactions. Eleven countries had indicated their interest in going ahead: Austria, Belgium, Estonia, France, Germany, Greece, Italy, Portugal, Slovakia, Slovenia and Spain. For Rio+20, the idea receded into the long grass as the eurozone focused on the rescue packages for Greece, Ireland, Portugal and Spain. There were some other developments that were important for addressing future financial flows. This included the G77's proposal that an intergovernmental process under the UNGA be set up to "propose options on an effective Sustainable Development Financing Strategy." This approach is similar to discussions at Rio+5 that became the "Financing for Development" process. "The Future We Want," finally, enshrines the creation of an intergovernmental panel on resource mobilization, which will be critical to putting together the funding options for the post-2015 Sustainable Development Goals.

### *A technology facilitation mechanism*

Several paragraphs of the Rio+20 Outcome Document address the crucial subject of technology transfer. For example, paragraph 273 requests:

> relevant United Nations agencies to identify options for a facilitation mechanism that promotes the development, transfer and dissemination of clean and environmentally sound technologies by, inter alia, assessing the technology needs of developing countries, options to address those needs and capacity-building [and] requests the Secretary-General, on the basis of the options identified and taking into account existing models, to make recommendations regarding the facilitation mechanism to the General Assembly at its sixty-seventh session.
>
> (United Nations, 2012)

A technology facilitation mechanism could input suggestions for the technical aspect of the Sustainable Development Goals being developed and would promote four types of global and regional networks, building on existing ones. These networks would have a firm science foundation and bring in a network of business incubators and policy networks. The global network of national business incubators would, it is proposed, support business plans and product development; it would help build capacity for production skills and provide seed money. The network would aim to link together entrepreneurs and start-ups, pro-poor community enterprises, and various UN and non-UN support programmes.

## References


Bentley University (2012) Jean-Christophe Carteron, Available online at: www.bentley.edu/events/symposium/speakers/jean-christophe-carteron

Black, R. (2012) Rio Revisited: Glass Half-Full? Available online at: www.bbc.co.uk/news/science-environment-18967011

Chile (2012) Note verbale dated 27 June 2012 from the Permanent Mission of Chile to the United Nations addressed to the Secretary-General of the United Nations Conference on Sustainable Development, A/Conf.216/13, New York, UN. Available online at: www.eclac.org/rio20/noticias/paginas/8/48588/Declaracion-eng-N1244043.pdf

Corporate Sustainability Reporting Coalition (2013) Letter to the UK prime minister on corporate reporting, Norwich, Aviva. Available online at: www.aviva.com/media/news/item/the-eu-in-2013-embedding-corporate-sustainability-reporting-15615/

Dodds, F. and Nayer, A. (2012) Rio+20: A New Beginning, *Perspectives*, issue no. 8, Nairobi, UNEP. Available online at: www.unep.org/civil-society/Portals/24105/documents/perspectives/ENVIRONMENT_PAPERS_DISCUSSION_8.pdf

Euromed Management (2012a) Commitment to Sustainable Practices of Higher Education Institutions on the Occasion of the United Nations Conference on Sustainable Development. Available online at: http://rio20.euromed-management.com/Declaration-for-HEI.pdf

Euromed Management (2012b) List of Universities in HESI. Available online at: http://rio20.euromed-management.com/roll-of-commitments-2/

Euromed Management (2012c) Organizations Endorsing HES. Available online at: http://rio20.euromed-management.com/endorsers/

Foti, J. (2012) Beyond Rhetoric at Rio+20 with the "Compendium of Commitments," *WRI Insights*. Available online at: http://insights.wri.org/news/2012/05/beyond-rhetoric-rio20-compendium-commitments

ICSU (2012) Future Earth: New Global Platform for Sustainability Research Launched at Rio+20, press release, Paris, ICSU. Available online at: http://www.icsu.org/news-centre/press-releases/press-releases-2012/future-earth-new-global-platform-for-sustainability-research-launched-at-rio-20/download-press-release

IISD (2012a) Business Action for Sustainable Development (BASD) 2012 Business Day. Available online at: www.iisd.ca/uncsd/rio20/basd2012/

International Chamber of Commerce (2012) International Chamber of Commerce Releases Roadmap for Green Growth. Available online at: www.iccwbo.org/News/Articles/2012/International-Chamber-of-Commerce-releases-Roadmap-for-Green-Growth/

Local Authorities Major Group (2013) Agenda 21: Chapter 28, New York, UN, 15 June 2013, Available online at: www.uncsd2012.org/index.php?menu=101

Mitchell, A. (2012) Andrew Mitchell Flags Up UK's 0.7% Overseas Aid Target, London, BBC News. Available online at: www.bbc.co.uk/news/uk-politics-18568533

NRDC (2012) The Rio+20 Earth Summit: Actions for Accountability. Available online at: www.nrdc.org/international/rio-2012/accountability.asp

Ramstein, C. (2012) Voluntary Commitments: Delivering Promises on Sustainable Development? Available online at: www.iddri.org/Publications/Collections/Idees-pour-le-debat/WP2312_CR_Rio+20%20voluntary%20commitments.pdf

SIDS Energy for All (2012) Small Island Developing States Sustainable Energy for All Conference: What Is SIDS SE4All? Available online at: www.sidsenergyforall.org/conference-details/what-is-the-conference-about/

Steven, D. and Evans, A. (2012) Open Letter to the Co-Chairs of the UN High Level Panel on the Post-2015 Agenda, Available online at: www.globaldashboard.org/2012/05/14/open-letter-to-the-co-chairs-of-the-un-high-level-panel-on-the-post-2015-agenda/

Thompson, H. E. (2012) Letter on the Higher Education Initiative. Available online at: www.planetworkshops.org/en/post/1042/the-higher-education-initiative.html

UNCCD (n.d.) National Councils for Sustainable Development (NCSDs), Bonn, UNCCD. Available online at: www.unccd.int/en/programmes/Capacity-building/CBW/Features/Pages/National-Councils-for-Sustainable-Development-(NCSDs)-.aspx

UN DESA (2012a) Higher Education Sustainability Initiative for Rio+20. Available online at: www.uncsd2012.org/index.php?menu=163

UN DESA (2012b) Rio+20's Compendium of Voluntary Commitments. Available online at: www.uncsd2012.org/voluntarycommitments.html

UN DESA (2012c) Rio+20 The Future We Want, New York, UN. Available online at: http://sustainabledevelopment.un.org/futurewewant.html

UNEP (2012a) Natural Capital Declaration. Available online at: http://www.naturalcapitaldeclaration.org

UNEP (2012b) Rio+20 Declaration on Justice, Governance and Law for Environmental Sustainability, Nairobi, UNEP. Available online at: www.unep.org/rio20/Portals/24180/Rio20_Declaration_on_Justice_Gov_n_Law_4_Env_Sustainability.pdf

United Nations (1992) Agenda 21, New York, UN. Available online at: http://sustainabledevelopment.un.org/content/documents/Agenda21.pdf

United Nations (2012c) The Future We Want, New York, UN. Available online at: www.un.org/ga/search/view_doc.asp?symbol=A/RES/66/288&Lang=E

United Nations Global Compact (2012) Rio+20 Corporate Sustainability Forum, Available online at: http://csf.compact4rio.org/events/rio-20-corporate-sustainability-forum/custom-121-251b87a2deaa4e56a3e00ca1d66e5bfd.aspx

United Nations Visitors Centre (2012) Jonas Haertle. Available online at: www.un.org/wcm/content/site/visitors/lang/en/home/to_see_and_do/lectures/alpha_list_h

World Bank (2012a) Massive Show of Support for Action on Natural Capital Accounting at Rio Summit. Available online at: www.worldbank.org/en/news/press-release/2012/06/20/massive-show-support-action-natural-capital-accounting-rio-summit

World Bank (2012b) Rio+20: Natural Capital Accounting and the Wealth of Countries. Available online at: www.worldbank.org/en/news/feature/2012/05/30/rio-20-natural-capital-accounting-feature

Chapter 5

# From Rio+20 to 2015 and the new development agenda

## Was Rio+20 a failure or a success?

The press corps in Rio, and in comments made since, was almost unanimous in claiming that Rio+20 had been a global disaster. Yet even within days that view was being revised. One of its key critics, John Vidal of the *Guardian*, looked back as early as a week after world leaders had left and the dust had settled over Rio Centro:

> Reaction after the 1992 Rio summit was uncannily similar. Countries passed then what now seem far-sighted treaties and embedded a slew of aspirations and commitments into international documents – but NGOs and journalists were still distraught. They said the climate change agreement was too weak, that sustainable development was too abstract a concept, that the promised aid was inadequate, and that the US had guaranteed the felling of the Amazon forest by refusing to sign the biodiversity convention. There were, they said, no agreements on population growth or subsidies, or oceans, or trade, or women's rights . . . and myriad other issues. In short, just like Rio 2012, the meeting was said [in 1992] to be a dismal failure of governments to co-operate.
>
> (Vidal, 2012)

Cynics will always say that the conference was ultimately all about face-saving, believing that all the hard compromises, the tough choices concerning our future, were avoided in pursuit of a consensual result. This criticism will never be fully dispelled. Yet, a macro-conference of the magnitude of Rio+20 will be measured not only by what it delivered on paper but also by the transformations it entailed and will ultimately engender. More than a point of arrival where many processes were meant to converge, Rio+20 was consciously built to be a point of departure for a journey compelling governments, corporations, organizations and, ultimately, individuals to find ways to live, thrive and equally prosper within the finite social and environmental thresholds imposed by our planet. From this perspective, others will look back – in 2016, in 2030 and eventually in 2050 – and will judge whether the seeds that we planted in 2012 in Rio de Janeiro meant something.

Rio+20 was built on the notion that much had happened in the world since 1992 when Agenda 21 was adopted. Chapter 1 of this book stressed the idea that during 2000 and 2012, several trends led towards a strange rebirth of sustainable development. From being a notion that at the beginning of this century had been abandoned to the sidelines of our global political discourse, sustainable development came back as the coalescing leitmotiv for our collective aspirations for a better and more equal life for us and for our daughters and sons. That same year, 2012, we also initiated a complete reassessment of the way the United Nations had dealt with our present (eradicating poverty and reducing inequalities) and will address and chart our future.

As Rio+20 became old news, a process was nevertheless already under way to reassess to what extent our current narratives and discourses on sustainable development were still valid. To this end, "The Future We Want" left us some signposts. First, we need to build a compelling and inspiring narrative for sustainability, since facts alone are not enough to trigger action. Second, there is a need to reassess the global development goals and objectives under an integrative development agenda. The third imperative was to assess our technological means and financial resources for a new partnership for development. Fourth, we need to review the role of institutions for sustainable development. Finally, it is necessary to open the door fully to major groups and other stakeholders, since it is now clear that change, and sustainable change, cannot happen without a partnership between stakeholders and governments.

Despite the overwhelming challenges that the world faces, we are convinced that today we are in a better position to do the work than we ever were in the past two decades. There are several reasons for our cautious optimism. The gradual assimilation of sustainable development as our only possible standard for long-term prosperity is, for starters, a major change from the situation in 2000, when gross national product (GNP) was thought to be the gold standard of progress. The second is the increasing influence, social and economic progress and participation of a larger, more representative group of developing countries. With greater influence should come also greater responsibilities. In the coming years we will see a greater emphasis on redefining the role of the middle-income countries in the implementation of a new development agenda.

For the United Nations and the multilateral system, this is also a moment of redefinition. It is fair to say that on the eve of the year 2000 the United Nations was unprepared to deliver and ensure the follow-up of its own goals set out in the Millennium Declaration. It took two years for the General Assembly to endorse a basic set of targets to measure the implementation of the MDGs and to agree on the Monterrey Consensus, which is our international framework for coordinating the international efforts for financing for development.

It was not until 2005, five years after the MDGs had been agreed, at the first comprehensive review of the Millennium Declaration – also known as the

World Summit Outcome – that the United Nations spearheaded the establishment of the basic tools for supporting the implementation and follow-up of the MDGs. These were the Economic and Social Council's (ECOSOC) Annual Ministerial Review (AMR) and the Development Cooperation Forum (DCF). The World Summit Outcome initiated a series of reforms of the operational activities for development aimed towards a more effective and accountable UN country presence, and also started a number of other institutional processes on the rule of law and the international environmental governance. With the notable exceptions of the creation of the Human Rights Council and the Peacebuilding Commission, the 2005 reform only tinkered on the margins of the major systemic weaknesses. Among these were the reform of the governance, surveillance and response functions of international financial institutions; the fulfilment by the pledged date of January 2005 of the trade negotiations under the Doha Development Round; the reform of the organs of the United Nations (General Assembly, Security Council and the Economic and Social Council); and the fragmentation of the treaty bodies dealing with the environment. Although we do not claim any causality, we believe that the failure to address these structural weaknesses left the United Nations and the multilateral system unprepared to frame and guide the global recovery during and after the international financial and economic crises of 2008.

The major difference that we see today is the mapping of the major trends – demographic, economic and environmental – that will define during the next three decades the future of global efforts to achieve sustainable development.

The second difference is that the United Nations and the multilateral system, our own institutions, today seem more prepared to rise to the challenge of supporting a global sustainable development agenda. With more than two years of anticipation, Rio+20 set an intergovernmental process for defining global Sustainable Development Goals as an integral part of the Post-2015 Development Agenda. Building on the experience of almost 15 years of implementation of the MDGs, UN member states have aligned the institutional framework before setting the goals. This will undoubtedly improve the basis for following up and reviewing a new development agenda. The institutional architecture, of which several key parts have already been established, is attempting to recover the centrality of the United Nations as the norm-setting forum for development. It also strives, drawing from the deficiencies of the Commission of Sustainable Development (CSD), to provide a clear differentiation between sustainable development and environmental policy-setting. The reform of multilateral institutions in charge of development is an urgent matter. We must integrate a diversity of isolated processes around the United Nations' development agenda, benefiting from synergies and avoiding the scattering of already scarce resources, within a renewed strategic framework where each of the relevant stakeholders fulfils its corresponding role and responsibility. The United Nations must not have parallel development agendas.

The third factor is the unprecedented commitment from the World Bank, the International Monetary Fund and the World Trade Organization to make sustainable development their own business standards. The final factor is the deadlines set by UN member states to address, in the next three years, other major structural components in concluding the ongoing international trade negotiations and defining by 2015 a new universal legal regime for addressing climate change.

This book has been a journey taking the reader from the road to Rio+20 to Rio itself. We hope it has explained why certain outcomes were possible and why others were not. This chapter will consider what has happened since Rio+20, the state of the processes that were initiated, and how they are merging with the preparations for the final assessment of the Millennium Development Goals, which together are referred to as "the Post-2015 Development Agenda." It will cover the development of a new narrative for advocating for one single sustainable development framework with one set of universal Sustainable Development Goals; it will also touch on the role of institutions for sustainable development. Finally, it will end with a look at where we should be on January 1, 2016, once the convergence of these processes is concluded.

## A new, inspiring narrative: development for all within the limits of our planet

While the principles that underpin sustainable development have remained static through the past decades, its conceptual interpretation has been quite fluid. An overview of the different ways that the concept of sustainable development has been referred to in intergovernmental documents in the past two decades shows us that changes are possible, yet are never made overnight. They take time. They require determined efforts and compromises.

The Brundtland Commission's 1987 vision of sustainable development was a notion eminently based on ethical and moral considerations by emphasizing the imperative of intergenerational solidarity. Now, nearly three decades on, it seems that sustainable development is based on complementing increasing standards of well-being with the life-supporting functions of the planet.

For this vision, the policy implications differ from recommending to decision-makers, as the Brundtland Report did, a harmonious balancing of social development, economic development and environmental protection policies. A new narrative of sustainable development will necessarily be grounded in making explicit a global commitment to ensure the right to development for all, but within the limits established by widely known environmental thresholds. It must also move away from being seen as synonymous with an environmental agenda.

After Rio+20 there was a need to bring forward an ambitious and inspiring narrative for sustainable development to make it a core component of the UN development agenda, thereby ensuring that sustainability becomes the operating ethic for government, business and society. Ban Ki-moon, the UN

Secretary-General, having made sustainable development a priority, was among the first leaders to understand this need. In his statement at the opening of the Rio+20 Conference, he referred to the agreement to establish a process to define universal sustainable development goals set out in paragraph 248 of "The Future We Want." He was adamant he would "spare no effort to implement the mandate given by the Member States to realize our vision of sustainable development goals that build on the success of the MDGs" (United Nations, 2012a).

Shortly afterwards, he announced the launching, under his authority, of two major interrelated initiatives aimed at building bridges between Rio+20 and the follow-up to the MDGs: the Sustainable Development Solutions Network and the High-Level Panel of Eminent Persons on the Post-2015 Development Agenda. Both initiatives were intended to provide the United Nations and its member states with a basis, building on "The Future We Want" and the Millennium Declaration, on which to construct a narrative (or, in policy terms, a "framework") for global sustainable development that will enact our global sustainable development agenda once the MDGs' initial term comes to a conclusion in 2015.

While both of these initiatives aim at strengthening the intergovernmental policy definition processes, their scope and background are different. The Post-2015 High-Level Panel is grounded in the thrust and political commitment provided by world leaders and eminent personalities whose work was preceded by national stakeholder consultations in over 100 countries and eleven thematic consultations. The Sustainable Development Solutions Network (SDSN), set up by Jeffrey Sachs under the instruction of the UN Secretary-General, is, on the other hand, an epistemological endeavour. It consists of an independent global network of research centres, universities and technical institutions whose function is to provide scientifically credible and policy-relevant advice to governments and NGOs, including business, other stakeholders, UN agencies and other international organizations involved in the crafting of a new development agenda (United Nations, 2012d).

The SDSN proposed that sustainable development be defined as "a holistic concept addressing four dimensions of society: economic development (including the end of extreme poverty), social inclusion, environmental sustainability and good governance including peace and security" (SDSN, 2013). For the SDSN there was also a need to further depart from the 1987 "Brundtland vision," for which sustainable development was above all a translation of intergenerational equity and solidarity: ensuring the needs of present generations without compromising those of future generations. The SDSN application of the paradigm of sustainable development was grounded in the growing amount of scientific evidence accumulated during the past decades on poverty as a multidimensional phenomenon, as well as indisputable scientific evidence on the impacts of human activities on land, water, the oceans and atmospheric resources, which was not available in 1987 and was highlighted in work on planetary boundaries. This new notion of mutual reliance (right to development and limits of the planet) is otherwise elaborated in the SDSN proposal – which draws from the chapter on scenarios of the 2012 *Global Environment Outlook* (*GEO-5*; UNEP, 2012) in the form of trajectories and failures: failures in environmental sustainability or gender equality can undermine progress in other areas, such as eradication of poverty. By the same token, and drawing from the lessons of the implementation of the MDGs, poor governance and insecurity can easily undermine progress on economic, social, and environmental objectives. A sustainable development framework under this new logic requires important normative and behavioural shifts at all levels. The SDSN report argues for four normative concepts:

1 the right to development
2 human rights and social inclusion, allowing all individuals to have equal opportunities, to be treated equally before the law and to have equal access to public services;
3 convergences as all regions continue to raise living standards, with poor countries experiencing higher growth rates;
4 shared responsibilities and opportunities, while giving low-income countries and those with special vulnerabilities the international support they need to end extreme poverty and access the technology needed, and accounting for higher responsibilities for current and historic per capita use of environmental resources by high-income countries.

Less than a year after its launching, the Leadership Council of the SDSN initiative – a consortium composed of over 60 highly recognized leaders in the academic, political and business fields – submitted to the secretary-general a report entitled *An Action Agenda for Sustainable Development* (SDSN, 2013). It identified ten priority challenges for sustainable development:

1 ending extreme poverty and hunger;
2 achieving development and prosperity for all, within planetary boundaries;
3 ensuring effective learning for all children and youth for life and livelihood;
4 achieving gender equality, social inclusion and human rights for all;
5 achieving health and well-being at all ages;
6 improving agriculture systems and raising rural prosperity;
7 making cities inclusive, productive and resilient;
8 curbing human-induced climate change and ensuring sustainable energy;
9 securing ecosystem services and biodiversity and ensuring sound management of water and other natural resources;
10 transforming governance for sustainable development, including by improving corporate governance and aligning business behaviour with all the goals.

Several factors militated against a smooth landing for the SDSN's contribution. The development of a set of SDG suggestions by the SDSN seemed to many to go beyond the initial role envisaged for the Network and risked diverting attention from the High-Level Panel on the Post-2015 Development Agenda's own list of suggested goals and targets. The overlapping timings for the presentation of the two reports made also difficult a proper assimilation and discussion by UN member states of the SDSN contribution. In addition to this, it goes without saying that the reinterpretation of the dimensions of sustainable development encompassing governance and the rule of law stirred further controversy, despite being one of the major achievements of the 2012 High-Level Meeting of the UNGA on the rule of law, in which linkages were set between sustainable development and the rule of law. The process for building the SDSN report also ended up causing another set of concerns among stakeholders, as they had their own processes set up to bring in different stakeholders' views. In an attempt to address this concern, the SDSN started a wider consultation prior to publishing its final report. Many stakeholders took part, some reluctantly, but taking the view that submitting their views to any and all avenues for influencing the development of the new post-2015 development goals was worth it. All these factors should not shadow the significant contribution made by the SDSN in signposting the scientific evidence that a new development compact must be founded in the idea that sustainable development entails a mutual reliance between the right to development for all and the limits of the planet.

The High-Level Panel of Eminent Persons on the Post-2015 Development Agenda was launched only a few weeks after the closing of Rio+20. In it, Ban Ki-moon requested a group of "eminent persons" to "prepare a bold yet practical development vision to present to Member States in 2013." In contrast with the SDSN, which was in itself an innovative platform, the structure and organization of the High-Level Panel were similar to those of previous high-level endeavours, the latest being the Global Sustainability Panel. The panel consisted of a cross-cutting selection of 26 personalities drawn from government, the private sector, academia, civil society and youth. Three serving leaders, whose personalities and policies embodied to a great extent the transformational role that development plays, agreed to preside over its work: President Susilo Bambang Yudhoyono of Indonesia, President Ellen Johnson Sirleaf of Liberia and Prime Minister David Cameron of the United Kingdom. The High-Level Panel concluded its work at the end of May 2013 and presented to the Secretary-General and the UN member states its own report, entitled *A New Global Partnership: Eradicate Poverty and Transform Economies through Sustainable Development.*

The Post-2015 panel built on other UN commissions and panels by engaging in outreach to stakeholders around the planet. It differed in one significant aspect: its ability to utilize the new level of social outreach provided by social media and a combination of massive and targeted global, national regional and thematic consultations.

The idea of convening a panel on what would come once the MDGs expired was not entirely new. Initially, it drew from an intergovernmental mandate given in 2010 to the Secretary-General during the MDG review summit in which UN member states asked him to "report annually on progress in the implementation of the Millennium Development Goals until 2015 and to make recommendations in his annual reports, as appropriate, for further steps to advance the United Nations development agenda beyond 2015" (United Nations, 2010). The panel had thus been in the making for several months prior to Rio+20, as became clear after it became known in early April 2012 that the United Kingdom's prime minister had agreed to co-lead an effort on "a new UN committee tasked with establishing a new set of UN millennium development goals to follow the present goals, which expire in 2015" (*Guardian* (London), April 12, 2012). The original scope of proposal for a panel was substantially broadened after Rio+20.

Rio+20 and the agreement on the SDGs sank any limited and focused approach on development understood as poverty eradication. This led Ban Ki-moon to pose, in July 2013, a challenging and unexpected endeavour to its panel members: "I look forward to the recommendations on a global post-2015 agenda with shared responsibilities for all countries and with the fight against poverty and sustainable development at its core" (Ban Ki-moon, 2012). Instead of drawing only from the MDGs, as was originally expected, the post-2015 panel members were asked to prepare their report taking equally into consideration the Millennium Declaration and "The Future We Want," thus forcing, from the highest executive office of the United Nations, a complete reassessment of the United Nations' developmental pillar. To support this decision, a UN system-wide report entitled *Realizing the Future We Want for All* (United Nations, 2012b) played a fundamental role. This report was prepared by the UN System Task Team, established in September 2011 and composed of senior experts from over 50 UN entities and international organizations led by the Department of Economic and Social Affairs (UN DESA) and the United Nations Development Programme (UNDP). The final recommendations of the report had to be substantially redrafted in light of the results of Rio+20, illustrating the extent of the impact that Rio+20 had on the strategic thinking of the UN organizations. Its main recommendations were as follows:

- The central challenge of the Post-2015 Development Agenda is to ensure that globalization becomes a positive force for all the world's peoples of present and future generations.
- There is a need for a vision of the future that rests on the core values of human rights, equality and sustainability.
- There is a need for a more holistic development approach, grounded on key strengths of the MDGs but encompassing inclusive social development, inclusive economic development, environmental sustainability, and peace and security.

- The Post-2015 Development Agenda should be conceived as a truly global agenda with shared responsibilities for all countries, thus redefining the global partnership for development.
- The vision and framework for the Post-2015 Development Agenda has to be fully aligned with "The Future We Want" and be developed on the basis of broad and inclusive consultation processes.

The Post-2015 panel benefited from past experiences. A first improvement was securing the level of internal leadership by entrusting the Deputy Secretary-General, Ambassador Jan Eliasson, with the task of overseeing the post-2015 process. The working methodologies were also reviewed by establishing a dedicated and independent secretariat headed by a lead author, Homi Kharas from the Brookings Institution, someone more versed in the workings of the World Bank than in those of the United Nations. The risk of a Sherpa-driven process was also avoided by allowing the co-chairs and the panel members to take ownership of the process right from the start. Finally, early timing also played a role in helping UN member states and stakeholders at large to take stock of the report, as the panel concluded its work in May 2013 at a moment when no agreement on a road map to the post-2015 agenda was yet in sight and when only two meetings of the Open Working Group on Sustainable Development Goals (OWG) had taken place, owing to the late launching of the major follow-up process of Rio+20.

The key message by the panel, "one world: one sustainable development agenda," and the five transformative changes highlighted in its report build on all of the sections of the 2000 Millennium Declaration and on those systemic trends already identified during the previous year by the Global Sustainability Panel (United Nations, 2012), the UN System Task Team (United Nations, 2012) and the Committee on Development Policy (United Nations, 2012). The five transformative shifts that the panel highlighted (United Nations, 2013) were:

1. *Leave no one behind.* This includes tackling the causes of poverty, exclusion and inequality.
2. *Put sustainable development at the core.* This entails that all development models have to include environmental thresholds and set out targets for sustainable lifestyles for all (patterns of consumption and production).
3. *Transform economies for jobs and inclusive growth.* This entails the creation of opportunities, of physical and social infrastructure, for good and decent jobs, allowing a massive global progression in poverty levels over the coming decades.
4. *Build peace and effective, open and accountable institutions.* This entails building transparent, accountable and responsive governments and institutions that respect basic internationally recognized rights such as access to justice, freedom from discrimination and persecution, and involvement and participation in decision-making processes.

5 *Forge a new global partnership.* This entails merging the people-centred and the planet-centred agendas, crafting a clear method of measuring universal progress through goals and targets, and basing the future development framework on the basis of basic principles contained in the UN Charter, the 1966 International Covenant on Economic, Social and Cultural Rights, the 1992 Rio Principles on Environment and Development and the Millennium Declaration.

We highlight three key contributions made by the High-Level Panel. The first was building a political platform for advocating for one single sustainable development framework with one set of universal Sustainable Development Goals. The second was to give impetus to the UN system to launch the global conversation on a new development agenda, which at its height in 2013 gave a say to more than 800,000 participants in 194 countries though the social media campaign MyWorld2015. The third substantive innovation brought forward by the High-Level Panel (United Nations, 2013) lay in its summarizing of its vision in a limited list of 12 illustrative universal Sustainable Development Goals and 54 national targets:

1 end poverty;
2 empower girls and women and achieve gender equality;
3 provide quality education and lifelong learning;
4 ensure healthy lives;
5 ensure food security and good nutrition;
6 achieve universal access to water and sanitation;
7 secure sustainable energy;
8 create jobs, sustainable livelihoods and equitable growth;
9 manage natural resource assets sustainably;
10 ensure good governance and effective institutions;
11 ensure peaceful societies;
12 create a global enabling environment and catalyse long-term finance.

## Institutions for a new development agenda

An integrated development agenda requires coherent and responsive institutions at all levels. Institutions are more than organizations; they span our norms, rules and principles that govern our interactions. International development goals and objectives form part of our institutionalized norms. A constant concern of this book has been to highlight that our global institutional framework is in transition. Chapters 1 and 2 stress that the international norms and principles that define the common aspiration for a safer, healthier life and equal standards of living, which together form the foundation of the United Nations, have seen a complex process of convergence since 2000. This convergence was initiated with the adoption of a holistic human-centred framework in the Millennium

Declaration and in an accelerated manner after 2008, when the international financial crisis and the emergence of new actors led to a "strange rebirth of sustainable development."

Institutions are important mechanisms for delivering the development agenda. The discussions around reform to date have not sufficiently sought to align form and function of institution with a new development agenda. While much focus has been placed on new institutions and structures within the UN system, little attention recently has been paid to the reforming of existing structures such as the Charter body the Economic and Social Council (ECOSOC), UN programmes such as the United Nations Development Programme (UNDP) and secretariat bodies such as the United Nations Department for Economic and Social Affairs. The reform of these and the collaborations between them and other bodies will become necessary and are central to delivering on the sustainable development agenda. Moreover, discussion will now have to take place on the national and regional institutional architectures that will be most effective for translating sustainable development from an international to a national agenda.

This shift at the national level can only take place if presidents and prime ministers, ministers of finance, ministers of economic affairs and those with the greatest influence in cabinets embrace sustainability as central to economic and social development, and not merely an environmental issue. International meetings on sustainability must encompass multidisciplinary and multi-sectoral approaches, and bring together actors outside of the environmental sphere.

## How can universal Sustainable Development Goals build a single yet differentiated development agenda?

The Sustainable Development Goals (SDGs) were originally proposed at a government workshop for Rio+20 in July 2011 in Solo, Indonesia (see Chapter 2). The discussion on the SDGs changed the focus of the Post-2015 Development Agenda. Prior to Solo, the approach to any new goals to be agreed during Rio+20 was focused mainly on revitalizing existing international goals contained in the 2002 Johannesburg Plan of Implementation in areas such as oceans and marine biodiversity, waste, and management of chemical products, and incorporating new areas that had been overlooked by the MDG framework – Sustainable Energy for All being a prime example. After Solo and during Rio+20 the focus of the discussion turned towards the relationship between the SDGs and the MDGs and the identification of new priority areas, among which poverty eradication became the most often cited.

The final agreement reached in paragraph 248 of "The Future We Want," calling for the establishment of an intergovernmental committee – called the Open Working Group – to define during 2013 and the first half of 2014 Sustainable Development Goals "coherent with and integrated into the United

Nations development agenda beyond 2015," brought sustainable development to the centre of a new development goal discourse. It also broadened the discussion from measures addressed just to developing countries to ones addressed to all countries taking on the goals. These decisions opened a number of questions that were left unresolved in Rio, owing to the last-minute compromises that had to be made.

### *Agreeing on a transparent and inclusive approach*

The most immediate questions were ones of defining the leadership, composition and conduct of the activities of the Open Working Group (OWG). These issues often reflected differing views among countries. Some preferred a universal process under the General Assembly while others had actively advocated for a more restrained expert-driven process. There had been some opposition in the run-up to Rio to the SDGs becoming the main process for developing the new development goals, but since then countries have recognized that they do not want two parallel processes. The setting up of the OWG had a troubled birth because 70 countries wanted the 30 seats. This in the end caused some very interesting pairing up, breaking down traditional groupings, and has turned to be a source of strength (Evans and Steven, 2013).

Eventually, all of these questions were settled at the beginning of 2013, first with agreement during late January on the membership of the group, then with the definition of "methods of work" during the group's first session, which took place during mid-March 2013. The pressure built around the expectations of the outcome of the group was somehow initially relieved with the agreement that the group was meant to adopt its report and recommendations "by consensus, reflecting different options, if necessary." The leadership of the group was entrusted to two ambassadors who had previously played leading roles during the preparations for Rio+20 of the groups of friends of the institutional framework and water issues, respectively: Ambassador Macharia Kamau from Kenya and Ambassador Csaba Körösi from Hungary.

The group's work was organized into two main phases. The first phase, from March 2013 to February 2014, focused on taking stock: collecting views of experts, member states and other stakeholders. In the second phase, from February to September 2014, the group focused on preparing its recommendations in the form of a "strategic outlook" and a proposal for SDGs to the UNGA. From its beginnings, Ambassadors Kamau and Körösi affirmed their leadership by retaining the trust of all the members and non-members of the OWG and proposing a flexible and balanced agenda of issues.

The Group of 77 was particularly effective in encouraging the OWG to open its discussion to structural issues that had not been adequately addressed in "The Future We Want," such as the addressing of poverty as a multidimensional phenomenon; infrastructure development; and macroeconomic policy questions, including international trade; the international financial system; and

*Figure 5.1* Co-chairs of the Open Working Group on Sustainable Development Goals, Csaba Körösi, Hungarian ambassador, David O'Connor of UN DESA and Macharia Kamau, Kenyan ambassador (courtesy of IISD/*Earth Negotiations Bulletin* (www.iisd.ca)).

external debt sustainability. A greater emphasis was also placed on other social issues such as governance, the rule of law, conflict prevention, post-conflict peacebuilding and the promotion of durable peace. Such a scoping process right from the initial stages of the OWG reflected a need to go beyond the limitations set by the sectoral areas included in "The Future We Want," and proved in perspective that it had been a wise decision not to pre-define "priority areas" for the SDGs. It also allowed a smoother assimilation of the contributions and the report of the High-Level Panel of Eminent Persons on the Post-2015 Development Agenda.

**Box 5.1 Outline of the issues covered from March 2013 to February 2014 by the Open Working Group**

- Conceptualizing the SDGs.
- Poverty eradication.
- Food security and nutrition, sustainable agriculture.
- Desertification, land degradation and drought.
- Water and sanitation.
- Employment and decent work for all, social protection, youth, education and culture.
- Health, population dynamics.
- Sustained and inclusive economic growth, macroeconomic policy questions (including international trade, international financial system and external debt sustainability).
- Infrastructure development and industrialization.
- Energy.
- Needs of countries in special situations, African countries, LDCs, LLDCs, and SIDS as well as specific challenges facing the middle-income countries.
- Means of implementation (science and technology, knowledge-sharing and capacity-building).
- Global partnership for achieving sustainable development.
- Human rights, the right to development, global governance.
- Sustainable cities and human settlements, sustainable transport.
- Sustainable consumption and production (including chemicals and waste).
- Climate change and disaster risk reduction.
- Oceans and seas, forests and biodiversity.
- Promoting equality, including social equity, gender equality and women's empowerment.
- Conflict prevention, post-conflict peacebuilding and the promotion of durable peace.
- Rule of law and governance.

### *Combining universality and differentiation*

A second set of questions opened by the conceptualization of the SDGs was the need to define the universality of a new development agenda.

While the majority of countries expressed, early on, the view that the SDGs were an integral part of the ongoing discussions on the Post-2015 Development Agenda and expressed wide support for "a single post-2015 UN development framework containing a single set of goals – goals that are universally applicable to all countries but adaptable to different national realities and priorities," the issue of differentiation is one that cannot be avoided.

The financial crisis and its ensuing fallout in some of the more advanced economies, where it pushed even solidly middle-class families into poverty, served to awaken a new sense of global consciousness about the demeaning effect of poverty on the human person and spirit, and the fact that it compromises the optimal functioning of a society and economy, robs the country of the input of valuable human resources, causes a drain on national resources and diminishes the whole society. It resulted in a review of the entitlements of all citizens, not just the entitlements of the 1 per cent. Arising out of the financial crisis, the cry of the Arab Spring, the Occupy Wall Street movement and other social and economic protests and movements the world over had at their core the call for sustainable development, the desire of all people everywhere to live in dignity in this generation and the next, and to have a sustainable quality of life for the majority of people in every society the world over. The financial crisis ultimately catalysed a kinder global society where people want an end to poverty, access to justice, and a more equitable approach to the distribution of social, economic and environmental goods and services. It is this sentiment that now needs to be captured and replicated in national, regional, global and multilateral policies relating to the development agenda. Such an approach would have of necessity to be rights based.

## Resource mobilization and financing for sustainable development

In 1992, governments requested Maurice Strong, the secretary-general of the Earth Summit, to work out how much it would cost to implement Agenda 21. With the help of the UN agencies and programmes, he estimated that the costs would be $625 billion a year, with $125 billion going from North to South in development aid. The 1990s was a time of falling, not increasing, ODA, and it was only with the agreement of the Millennium Development Goals in 2000 that aid started to rise again substantively. The lack of delivery on the financial promises from 1992 had slowed down the implementation of Agenda 21.

The 1990s also saw net flows of private and commercial sources of funding increase fourfold. These large inflows of money were often concentrated in short-term debt, which had negative impacts when the flows were reversed. So, at Rio+20, developing country governments were keen to ensure that the same thing did not happen this time. Yet the compromise was made to set up another intergovernmental process, the "Committee on Financing Sustainable Development" (CFSD). This would work parallel to the SDG working group and produce a portfolio of options for financing the Sustainable Development Goals. The establishment of the CFSD will help the framework to be agreed upon in 2015 to be based in a clear idea of what the funding options for implementation will be. Its work must be to learn from what has and has not worked over the past 20 years. Between 2013 and 2015 we are likely to see revitalization of the follow-up process of the Monterrey and Doha Conferences on Financing for Development and the different spaces that ECOSOC provides to dialogue and define priorities with the international financial institutions.

### *Official development assistance commitments must be honoured*

Before Rio+20, the prevailing question among international cooperation actors that sponsored the idea of the post-2015 panel was about how to refine beyond 2015 a UN poverty eradication agenda by making it more effective and more focused, especially towards those countries whose peoples could not be left behind, including the least developed, those vulnerable to natural disasters and those emerging from post-conflict situations. This vision entailed a new approach on the modalities of the delivery of the development cooperation. Conceptually, however, this vision on a development agenda was still far from bridging sustainable development as an overarching umbrella for development cooperation. The announcement in 2011 of the United Kingdom's new aid policy is a positive illustrative example in this regard. This vision included, on the one hand, using international aid towards levelling the playing field for developing countries, particularly the least-developed ones, which included creating the "same conditions that create prosperity the world over," including access to markets, property rights and private-sector investment. On the other, it required improved effectiveness, drawing on partnerships involving the private sector and tapping into the increased potential of domestic resources mobilization. However, leadership for a new developmental vision also meant the need for donor countries to honour their commitments, particularly their targets on financing for development. As expressed by Prime Minister Cameron in 2011:

> I think there is a strong moral case for keeping our promises to the world's poorest and helping them even when we face challenges at home. When you make a promise to the poorest people in the world you should keep it.
> (Cameron, 2011)

Some of the financing for development commitments informing this vision included the Monterrey Consensus target of devoting 0.7 per cent of GDP on ODA, and the 2005 Gleneagles Commitment from the G8 and other donors to increase ODA to all developing countries by around $50 billion a year by 2010, compared with the 2004 figure.

A new development agenda must make sure that other types of commitments are kept, such as the commitment made in 2009 in Copenhagen to provide $30 billion in fast-start finance until 2012 and $100 billion per year by 2020 in long-term funds to support climate change action in the developing world, and the 2012 G20 Los Cabos climate change commitment, which included the full implementation of the outcomes of the Cancún and Durban climate change conferences and the support for the operationalization of the Green Climate Fund (United Nations, 2013).

### *New financing sources should also be identified*

The CFSD might take an approach whereby it offers different packages for different levels of development to fund and support the implementation of the new goals. The Human Development Index is a composite of life expectancy, education and income indices that rank countries into four tiers of development:

1 very high human development;
2 high human development;
3 medium human development;
4 low human development.

An approach that grouped these countries together on the basis of their level of development would enable a much more targeted discussion on the basket of resource and finance mobilization they would need to have available for funding the delivery of Sustainable Development Goals. This would also be an opportunity for learning from each other and learning what would be needed to help countries transform from one level to another.

In addition to targeted conversations on resource and finance mobilization, there would also need to be a discussion on "new financial mechanisms." The one most discussed over the past 20 years has been the financial transaction tax (FTT). This is one of the areas where there might be some movement. Eleven countries of the European Union are in the process of moving towards adopting a tax on financial transactions. One of the questions of course will be where the money is to be spent if the FTT succeeds. The United Kingdom has recently objected in the Court of Justice to the introduction of the tax even though it is not one of the 11 member states supporting it. Speaking at the International Monetary Fund's 2013 spring meetings, George Osborne, Britain's Chancellor of the Exchequer, said, "The British Government [is] not against financial transaction taxes in principle – we have stamp duty on shares – but we are concerned about the extra-territorial aspects of the Commission's proposal. That concern is shared by some other countries" (Aldrick, 2013).

Companies and individuals that have used tax havens in the past have contributed to weakening of the tax bases of states and their ability to fund public goods. At the same time, it should be considered that financial transactions are a significant source of funding for several small countries which are host to a number of offshore entities. Changes in the international regimes covering such entities and transactions will reduce the already reduced revenues of small-island governments, including governments in the Caribbean, further compromising their capacity to fund and sustain their development efforts.

An innovative new mechanism that is now available through the internet is crowd funding. Already used to fund films and music and many other things, crowd funding enables individuals or groups to fund something they like. These could as easily be development projects. NGOs could play a role as the facilitator for such a mechanism.

### *Make choices transparent; give a choice to investors and consumers*

Principle 10 of Agenda 21 opened the concept of transparency and the public's right of access to information relating to environmental issues. Increasingly, the push by the public is for less obscure, more open government processes, and equally for businesses to disclose information to consumers which allows them to make better-informed choices. As the agenda for sustainable consumption and production goes forward, these calls for readily available information will increase in number, frequency and intensity. It is only through access to information that the investor and consumer can protect themselves. Access to information is also consistent with the principles of corporate social responsibility, of which there are heightened levels and the opportunity in the future for a convention on corporate reporting. In what Peter Drucker has termed a "knowledge society" in which many jobs are knowledge and technology based, where technology puts information at the fingertips of the public within seconds, where the founder of Wikileaks and corporate whistleblowers are held in high regard by many, information – its capacity to spur multi-stakeholder dialogue and build trust, its potential in broadening understanding and accountability among important but (usually) separate actors, and its role in helping all actors to improve choices and decision-making – is a means of enlarging and enriching democracy.

The Declaration of the United Nations Conference on Human Development in Stockholm in 1972 contained a reference to the role of business in its preamble, calling for "the acceptance of responsibilities by citizens, communities and by enterprises and institutions at every level, all sharing equitably in common efforts." Subsequently, the Brundtland Commission, convened by the United Nations in 1983, set out to respond to a new generation of environmental problems such as climate change and deforestation, and combined the twin aims of environmental protection and poverty reduction into a single formula: sustainable development. The impact of international trade upon the environment was explicitly recognized in the Brundtland Report published in 1987, but primary responsibility for averting environmental degradation was ascribed to nation-states rather than corporations.

Regulation of corporations was again notable by its absence on the agenda of the 1992 United Nations Conference on Sustainable Development; although a chapter was prepared by the United Nations Centre on Transnational Corporations, it did not make its way into the final negotiations. The proposed chapter, entitled "Transnational Corporations and Sustainable Development," called for governments and international organizations to "address in future environmental instruments the rights and responsibilities of transnational corporations" and "establish or strengthen a regulatory environment supportive of sustainable development."

The role of corporations in sustainable development was explicitly tackled in Rio in 1992 in the Rio Declaration Principle 16 through its commitment to the "polluter pays" principle.

Calls for responsible investment and sustainability reporting have grown louder in the past 20 years. Parallel to this, a voluntary approach to companies producing sustainability reports was started up, called the Global Reporting Initiative (GRI). It has developed a comprehensive set of reporting guidelines and sector guidelines. Since the World Summit on Sustainable Development in 2002, as well as the growth of GRI there has also been the negotiation of ISO 2600 – the standard on social responsibility – and private-sector initiatives such as the Principles of Responsible Investment, not to mention the UN Global Compact. The banking and financial crisis that began in 2008 provided fresh impetus to debates around the role of corporations in delivering societal benefits.

At Rio+20 the push by a coalition of, primarily, finance institutions and NGOs and other stakeholders to secure an agreement for guidelines for companies listed on stock exchanges to report or explain why they were not producing sustainability reports did not ultimately succeed. Some of the key countries that supported it – Brazil, South Africa, France and Denmark – have continued to do so, as have the stakeholders in a loose coalition called the Group of Friends of Paragraph 47 – the paragraph dealing with the issue of reporting. Additional high-level support was given by the UN High-Level Panel, which said:

> We suggest that a mandatory "comply or explain" regime be phased in for all companies with a market capitalisation above $100 million equivalent.
>
> The same principle should apply to governments. National accounting for social and environmental effects should be mainstreamed by 2030.
>
> (United Nations, 2013)

Producing sustainability reports will make companies much more transparent to both the general public and investors. The companies of the future are those that already are minimizing their impacts on the environment and their use of natural resources. They are the leaders in the business community. Mandatory reporting will make it very clear which companies are part of the problem and we hope that the call by the High-Level Panel will be taken up by governments, but on a 2020 timeline, not a 2030 timeline.

At the global level a clear battle is shaping up between North and South on the issue of so-called tax havens. Starting about four decades ago, and with growing momentum within the past two, at a time of different international thinking, a number of double taxation treaties have been negotiated between countries in the developed world and a number of developing countries, largely in the Caribbean but also including British islands such as Jersey, Guernsey and the Isle of Man. These treaties allowed the development of vibrant offshore financial sectors which became a source of revenue for the beneficiary countries, where companies went to do business at a lower tax rate than that prevailing in their countries of origin. Some of the beneficiary countries were "no-tax" jurisdictions, while others were "low-tax" jurisdictions. The level of legislation, due diligence and regulation in the beneficiary country largely determined the extent to which offshore companies were "clean," engaging in business for lower tax and higher profits, or "dirty," engaging in business in offshore jurisdictions largely for the purpose of laundering money. In any event, the beneficiary country earned substantial revenues from the presence of these corporate entities within their territorial space, either from some level of taxes or from annual licensing and operating fees, the cost of operation of their commercial entities in the host country, related expenses, purchase of goods and services, as well as the living expenses of expatriate employees and their families while in the host country.

## A reformed Economic and Social Council and the High-Level Political Forum

One of the most surprising and encouraging results of Rio+20 was the abolition of the Commission on Sustainable Development (CSD). Chapters 2 and 3 identified some of the major weaknesses it had accumulated after two decades of existence. The abolition of the CSD, which held its last session on September 20, 2013, surprised many. Instead, "The Future We Want" committed to strengthen the Economic and Social Council (ECOSOC) as the principal organ in charge of the integrated and coordinated follow-up of the outcomes of all major conferences and summits of the United Nations in the economic, social, environmental and related fields, and recognized that ECOSOC contributes decisively to integrating the three dimensions of sustainable development in a balanced way. What are the motivations behind this move back to basics?

The coordinated follow-up of the outcomes of the major UN conferences in the economic, social, environmental and related fields, as well as in the definition of a unified, universal and coherent Post-2015 Development Agenda, is a major challenge. Such a challenge requires a reformed and strengthened

ECOSOC under a systemic approach encompassing the functional bodies, regional bodies, expert bodies, funds, programmes and other, related bodies. ECOSOC is the natural and established institution to address all global issues related to development. However, after several decades and poor performance, it is necessary to adopt radical measures to allow the Council, together with its system of Organic and Regional Commissions, to gain their effectiveness and capacity to define and adequately follow up a renewed development agenda, while involving all relevant stakeholders. In this context, ECOSOC's mission should be to generate the necessary common spaces for all relevant actors to engage in dialogue and make decisions.

The High-Level Political Forum established in paragraph 84 of "The Future We Want" and defined in the resolution A/RES/67/291 must act as a bridge that ensures convergence in the development agenda, not only between relevant actors at the global level but also within the United Nations, where coordination and complementarities between the General Assembly and ECOSOC are needed. The High-Level Political Forum is an "institutionalized process" to bring together all states, whether or not they are members of the United Nations, and other actors within and outside of the United Nations, including the Bretton Woods institutions, for policy definition, policy review, policy dialogue and cooperation for the implementation of an integrated sustainable development agenda. This section draws from the national submissions presented in 2012 to the Review of Resolutions 61/16 on the Strengthening of the Economic and Social Council (United Nations, 2012c).

In order to ensure the highest level of leadership, this space will be convened at the heads of state and government level every four years, in September, while ensuring annual high-level sessions with the participation of economic, financial, social, environmental and other ministers. The meetings of the forum will conclude with the adoption of a negotiated political document. To act as a bridge and to ensure that the United Nations works in a consistent and coherent manner, the forum will be convened on an annual basis by ECOSOC and every four years by the General Assembly. The functional, regional and expert bodies of ECOSOC will provide the machinery to support the implementation of the work of the High-Level Political Forum. The forum should not be perceived as an "organization," a "body" or an "institution"; it is devoid of governing structures and explicit membership, and it can only act through the UNGA and ECOSOC. This measure was decided with the aim of ensuring coherence in the governance of the sustainable development agenda.

The mission of the General Assembly as titular organ of the High-Level Political Forum will be to draw the necessary political leadership to ensure that the implementation of the Post-2015 Development Agenda remains a priority at the national and international levels. The Economic and Social Council (ECOSOC) should recover its original functions as spelled out in the Charter and contribute to identifying solutions and allocating resources in support of the priorities defined by the heads of state and government. Such challenges require a reformed and strengthened Economic and Social Council under a systemic approach encompassing the functional, regional and expert bodies, funds, programmes and other related bodies. We need to undertake a profound process of institutional reform that endows ECOSOC with the ability to fully achieve its mandate as defined in the Charter of the United Nations.

A process of reform must start from the premise that to have an effective Economic and Social Council, we need to equip it with a systemic vision allowing it to respond and act promptly upon the central challenges of development, while conferring value added on each of its instruments. A strengthened and reformed Economic and Social Council must have the capacity to fulfil five main functions, jointly with the High-Level Political Forum: seek integration of the three dimensions of sustainable development; support increased implementation at the national and global levels; strengthen effective system-wide coherence and coordination; involve all relevant actors and stakeholders; and strengthen the science–policy interface.

One of the major achievements of the High-Level Political Forum is the creation of a new set of tools for supporting increased and integrated implementation at the national and global levels. The forum, when organized "under the auspices" of ECOSOC, will conduct regular reviews, starting in 2016, on the follow-up and implementation of sustainable development commitments and objectives, including those related to the means of implementation. The development reviews will draw from the lessons of the CSD and the Annual Ministerial Reviews (AMRs), which will be replaced in 2016 once the Sustainable Development Goals are agreed. These new reviews will:

- be voluntary, while encouraging reporting, and will include developed and developing countries, as well as relevant UN entities;
- be state-led, involving ministerial and other relevant high-level participants;
- provide a platform for partnerships, including through the participation of major groups and other relevant stakeholders.

Another achievement of the High-Level Political Forum is the global sustainable development report. This is a recommendation originally articulated in the report of the Global Sustainability Panel (see p. 54). The forum is expected to consider, in 2014, the scope and methodology of a global sustainable development report, based on a proposal of the Secretary-General. The sustainable development report could be one of the main building blocks of a future global development compact, fully integrating sustainable development and poverty eradication as its overarching focus. It could fulfil some of the following functions (Laguna-Celis, 2012b):

- Provide a comprehensive, integrated and scientifically credible global sustainable development assessment to support decision-making processes at appropriate levels;

- Identify promising policy options to speed up achievement of the internationally agreed sustainable development goals . . . and enhance the implementation of sustainable development; . . .
- Strengthen the on-going process of capacity building for developing countries to conduct sustainable development monitoring and assessments;
- Inform the strategic directions of the relevant UN bodies and highlight relevant information from existing social, economic and environmental assessments.

## The United Nations Environment Assembly and a new governance model for environmental sustainability

While the compromise achieved in the Rio de Janeiro conference fell short of the expectations of many actors, Rio+20 contained the seeds for the transformation of the governance and capacities of the United Nations Environment Programme (UNEP) and provided opportunities to address in new and innovative ways the nexus between global policy definition in the field of environment and its integration to sustainable development. Paragraph 88 of "The Future We Want" and UNGA resolution 67/213 agreed to the upgrading of UNEP from a limited membership organization to one with universal membership. This has already happened, and the first meeting of the new, upgraded UNEP took place in February 2013. This strengthening of UNEP closed a long-standing chapter of the international environmental governance (IEG) debate (Laguna-Celis, 2012a).

The chapter that has been concluded was initiated two years before the World Summit on Sustainable Development 2002 (WSSD) when ministers of the environment gathered at the newly established Global Ministerial Environmental Forum (GMEF) invited the summit to review the requirements for a "greatly strengthened institutional structure at the international level." They presented for this purpose the outcome of an open-ended intergovernmental group known as the "Cartagena Package." At the WSSD in Johannesburg, delegations – while supportive of the full implementation of the package – fell short on deciding the establishment of universal membership for the Governing Council of UNEP. As a compromise, they requested the General Assembly to consider at its fifty-seventh session this "important but complex issue" (United Nations, 2002). However, despite the efforts undertaken in Nairobi and New York between 2003 and 2011, the question of how to build an inclusive, universal and legitimate environmental decision-making body within the United Nations remained open.

The decisions painstakingly negotiated in Rio and New York went beyond incremental reforms and have expanded and adapted to new circumstances the original mandate of UNEP as established in Resolution 2997 of 1972. These decisions also provide a strong platform to empower the United Nations Environment Assembly of the United Nations Environment Programme – as the governing body of UNEP is called – as the leading global environmental authority.

The major challenge now is implementation of the eight functions agreed to in paragraph 88 of "The Future We Want." This paragraph is essentially an institutional road map containing eight specific objectives and their corresponding functions, often providing indications on actions that need to be taken, while in other cases also requiring specific follow-up either in the framework of UNEP's governing body or through existing interagency coordination mechanisms.

At the outset it should be highlighted that paragraph 88 incorporates innovative governance and functional elements. It also draws from outcomes previously agreed at the ministerial level, in particular the Nairobi Declaration of February 7, 1997, the Malmö Ministerial Declaration of May 31, 2000 and the Nusa Dua Declaration of February 24, 2010. It borrows its structure from the objectives and functions identified in 2009 by the Rome–Belgrade outcome (UNEP, 2009). The measures contained in paragraph 88 were also influenced by non-papers and contributions presented in the weeks prior to Rio de Janeiro such as the G77 plus China's "UNEP final coordinator attempt" of May 2012, the Mexican non-paper "An Agenda for the Institutional Framework for Sustainable Development" of May 2012 and by paragraph 21 of the European Union's submission to the compilation document presented in November 2011.

The first Universal Session of the Governing Council of UNEP opened a new chapter in the process of reforming IEG. More than 150 delegations arrived in Nairobi prepared to set in motion a new governance model for UNEP, building upon its new mandate defined in paragraph 88 of "The Future We Want" and the UNGA resolution 67/213 adopted on December 21, 2012.

As a first step, ministers agreed, through their decision entitled "Implementation of paragraph 88 of 'The Future We Want,'" that all reforms aimed at strengthening the role of UNEP as "the leading global environmental authority that sets the global environmental agenda, promotes the coherent implementation of the environmental dimension of sustainable development within the United Nations system and serves as an authoritative advocate for the global environment." The overarching question was therefore how to build a new, inclusive and legitimate institution capable of fulfilling a renewed mandate. The first key decision was the recommendation extended to the UNGA to rename the Governing Council as the "United Nations Environment Assembly of the United Nations Environment Programme." The second decision was to abandon the existing arrangements for ministerial engagement within UNEP. The Global Ministerial Environmental Forum (GMEF), although useful as a universal space for policy dialogue, was deemed unfit to put ministers in charge.

In Nairobi, ministers agreed that each session of the United Nations Environment Assembly of UNEP will conclude with a two-day high-level segment as an integral part of its decision-making process. They also agreed that such a space should allow them to take strategic decisions and provide political guidance. The functions of this new high-level space will be to set the global environmental agenda; to provide overarching policy guidance and defining policy responses to address emerging environmental challenges; to review successful policies and exchange experiences; to engage with all relevant stakeholder dialogue, including through specific dialogues; and finally, to foster partnerships for achieving international environmental goals and for resource mobilization. The failure to ensure the active participation of all relevant stakeholders, particularly those from developing countries, was one of the major weaknesses of UNEP's old governance structure.

During the first universal session of the Governing Council of UNEP, ministers agreed to promote transparency and their effective engagement within UNEP's governance structures. To achieve this endeavour, all delegations committed to have ready by 2014 the following deliverables: to develop a process of modalities for stakeholder accreditation and participation that takes into account experiences from relevant UN bodies; to define mechanisms for receiving stakeholders' expert input and advice; and to review and adapt the working methods of the United Nations Environment Assembly of UNEP and the Committee of Permanent Representatives (CPR) to allow verbal and written contributions by all relevant stakeholders in the steps leading to the intergovernmental decision-making process. Finally, in response to a recommendation of the participants of UNEP's Civil Society Forum it was agreed to request the executive director of UNEP to establish a policy on access to written information and to report from 2014 to the United Nations Environment Assembly of UNEP on its implementation.

The promotion of a strong science policy interface was also considered to be a key role of the United Nations Environment Assembly of UNEP. As a response, ministers committed to identify by 2014 critical gaps in the interface and to enhance and improve the relevance of the "Summary for Policy Makers" of the *Global Environment Outlook* (*GEO*).

Concerning the day-to-day governance of UNEP, there was agreement to strengthen the role of the Nairobi-based CPR and it was decided to expand its current functions to provide advice on policy matters and to contribute to the preparation of the agenda of the United Nations Environment Assembly of UNEP; to prepare decisions for adoption and oversee their implementation; and to convene thematic and/or programmatic debates.

Two additional measures were adopted, to strengthen the role and accountability of the CPR. The first was to convene an open-ended meeting of the CPR, ensuring support to developing-countries representatives, especially those not resident in Nairobi, to provide advice and contribute to the preparation of the agenda of the United Nations Environment Assembly of UNEP. The second decision was to establish a subcommittee of the CPR that will meet annually to review and advance the preparation of key programmatic documents such as

UNEP's Medium-Term Strategy and its programme of work and budget in a manner coherent with the budgetary cycle of the United Nations. Besides these measures, the decisions adopted in Nairobi also included a road map aimed at expeditiously implementing the functions agreed to in Rio in other areas, such as the consolidation in Nairobi of the headquarters functions of the organization, the strengthening of UNEP's financing base, and its role and implementing capacities at the national and regional levels.

## The role of stakeholders in global policy-making

Rio+20 was another important step in the opening up of the United Nations to the involvement of stakeholders in its decision-making and the implementation of global agreements.

The Earth Summit in 1992 was the first UN conference for which governments had recognized that the involvement of stakeholders in the decision-making process secured better policy. The creation of the nine Major Group chapters of Agenda 21 reflected both the rights and the responsibilities of those stakeholders.

At the Five-Year Review of Agenda 21 in 1997, governments tried out, at the request of stakeholders, a more formal structure by which they could hear stakeholder views. This was through the introduction of multi-stakeholder dialogues. In the subsequent years, 1998–2001, the process developed into 12 hours of debating and informing governments about key issues under discussion at the UN Commission on Sustainable Development. The process also helped build greater trust and openness between governments and stakeholders, as well as influencing policy decisions.

The 2002 World Summit on Sustainable Development took the further step of agreeing that involving stakeholders in the implementation of policy agreements would make the latter more likely to be implemented. The launching of multi-stakeholder partnerships to help deliver the WSSD agreements, although these had mixed success, was a testing ground that has informed more recent attempts to develop multi-stakeholder partnerships.

By the time of Rio+20 a number of UN agencies and programmes were operating under a Major Group/stakeholder approach. There had been a steady move away from the "civil society" discourse that had dominated much of the 1990s to something which enabled a less diffuse process into a more targeted and stakeholder-specific approach. This respect for various voices enabled, for example, engagement with women about gender issues or local government on issues of implementation. Rio+20 took the next step: allowing stakeholders to input to the Zero Draft of "The Future We Want," thus involving stakeholders in the actual agenda-setting.

The past 20 years have seen the development of an increasing number of multi-stakeholder coalitions, some, though not all, including industry. These developments have again built stronger trust between different stakeholder groups. How these coalitions can be incorporated into any new process around the new development goals will be important. Clearly, each stakeholder group should have the chance to put forward its own views, but where such a group has arrived at similar positions to those of other stakeholder groups, the message that they can give to governments is strengthened.

As we approach 2015 and the agreement of new development goals, the United Nations and governments have been extensively reaching out to stakeholders to help define what these new goals, targets and indicators should be. They have been doing this through traditional ways such as national or global consultation as well as by utilizing new social media and in morning hearings and special intersessional meetings of stakeholders with the co-chairs or the Open Working Group on Sustainable Development Goals. This attempt at a more collective ownership for the new goals will, if successful, make them extremely powerful and create the possibility of a true partnership for development, or rather sustainable development, to be created and implemented.

In the coming years, this will put pressure on the Major Groups and other stakeholders operating around the United Nations to reform. There is wide recognition that they need to create much more transparency and accountability within each of these stakeholder groups.

Stakeholders will have to show that they are reaching out to their constituencies around the world. They will have to overcome the reality that it is often the larger, well-funded and traditionally Northern-based stakeholder organizations that are participating. Social media offer a great opportunity to reach out to constituencies that have often been inadequately represented. How those views are represented and integrated into new policy formulations will be one of the key challenges for the future. To this end, we support a review of the working methods of ECOSOC and the United Nations General Assembly to include in a proactive manner the contributions of different stakeholders according to their specific capacities and expertise, building on existing precedents from subsidiary bodies of ECOSOC such as the Committee on Food Security and the Commission of Sustainable Development. Such a review should be undertaken in a manner that is consistent with the mandate of ECOSOC and its NGO Committee as the "portal of entry" of major groups and other stakeholders. As we write this book, UN DESA's Division for Sustainable Development has recently brought out a report, *Strengthening Public Participation at the United Nations for Sustainable Development: Dialogue, Debate, Dissent, Deliberation*, which looks at the most progressive stakeholder engagements in the UN system and makes 17 recommendations (UN DESA, 2013).

## Ahead of January 2016

On January 1, 2016, as a new year begins, it will be much clearer whether the process that started in 2007 with President Lula's speech did in fact culminate in moving us towards "the future we want." The process set in motion on

Sustainable Development Goals, financing sustainable development, reform of UNEP and the new High-Level Political Forum will, we hope, have established firm foundations for a new development agenda and helped to build the trust to deliver it. To do so, further strengthening of the roles of civil society will have to take place, with emphasis on the private sector and a concerted effort to embed sustainable development as the underpinning philosophy in government policy and private-sector practices as well as the activities of individual practices and broad decision-making at all levels.

It is clear that the challenges ahead over the coming 20 years will be numerous. The real question is "Will we address these challenges together or in conflict with each other?" To help the reader, we have identified ten items that we believe could be agreed in 2015.

## *New development goals*

1 The 2015 conference agrees one set of new development goals for all countries.
2 Those goals could include ones such as:
   - ending extreme poverty;
   - achieving gender equality, social inclusion and human rights for all;
   - ensuring food security and good nutrition;
   - securing sustainable energy for all.
   - achieving universal availability of an acceptable quantity and quality of water for health, livelihoods, ecosystems, production and adequate sanitation;
   - achieving development while maintaining global temperature increase below 2 degrees;
   - creating decent jobs, sustainable livelihoods and growth that respects sustainability;
   - achieving health and well-being for all;
   - providing quality education and lifelong learning for all;
   - empowering the creation of inclusive, productive, resilient and sustainable urban areas.
3 With each goal there might be a set of targets that address different levels of development.
4 The indicators associated with each target should be developed using the frameworks developed by the UN Statistical Division.

## *Financing sustainable development*

To deliver a set of new development goals, the work of the UN Committee of Experts on Financing Sustainable Development will need to come up with a set of recommendations that include:

5 identifying the external mobilization of financial resources for sustainable development;

6 identifying innovative mechanisms for financing the protection of the global environment;
7 developing a matrix of policy options and financial instruments for countries to use depending on their level of development.

## *Climate change*

As well as agreeing a set of new Development Goals, 2015 will also see a new climate agreement. This should include the following points:

8 There must be a climate agreement for the period beyond 2020 that pledges to cut emissions from both developed and developing countries to make it possible to ensure that we do not go higher than a 2-degree increase in temperature.
9 Any agreement on a new climate regime must continue to take into consideration the principle of common but differentiated responsibilities.
10 Full funding for addressing climate change through the climate Green Fund is needed.

If these ten suggestions are in place on January 1, 2016, then we will know that Rio+20 did indeed create a new possibility for working together to create a more fair, equitable and sustainable planet.

Our final word is that this is all very important, but perhaps the time has come for the member states, stakeholders and the United Nations to agree on an "Earth Charter." The one developed by the Earth Council following Rio+20 with Maurice Strong and Mikhail Gorbachev at its head agreed to 4 pillars and 16 principles to guide this wonderful, diverse, more and more interdependent, and fragile world. Such an Earth Charter perhaps should be developed in a different way as compared with the traditional UN negotiations. We are all one human family and there is more that unites us than divides us. Perhaps January 1, 2016 will be the beginning of a global conversation that will culminate in 2020 in the adoption of an Earth Charter by parliaments (at all levels), companies, NGOs and other stakeholders. Let the last word go to Maurice Strong (1998):

> The Earth Charter is important as an expression of the commitment of people throughout the Earth to evoke their own deepest moral, spiritual and ethical principles in the task of ensuring a sustainable future for those who inhabit the Earth now and those who will follow us on the Earth.
>
> People act in accordance with their deepest motivations, and therefore to undertake major programs and activities, which we must do, they have to be rooted finally, in the priorities of people, and those priorities stem from their moral, ethical and spiritual values. So, the Earth Charter will attempt

to articulate these in rather universal terms. And by articulating them and ensuring that they represent the voice and the feelings of millions of people from all over the world, they can have an immense value in terms of motivating our political leaders to do what is necessary to ensure a sustainable future for our children.

## References


Aldrick, P. (2013) Britain Launches Legal Challenge to Financial Transaction Tax, *Daily Telegraph* (London), April 19. Available online at: www.telegraph.co.uk/finance/newsbysector/banksandfinance/10006501/Britain-launches-legal-challenge-to-Financial-Transaction-Tax.html

Ban Ki-moon (2012) Note to correspondents, July 31, New York, UN. Available online at: www.un.org/sg/offthecuff/?nid=2455

Cameron, D. (2011) Speech to the Global Alliance for Vaccines and Immunisation, June 13. Available online at: https://www.gov.uk/government/speeches/speech-at-vaccine-summit

Evans, A. and Steven, D. (2013) What Happens Now? The Post-2015 Agenda after the High-Level Panel, New York, Center on International Cooperation.

Laguna-Celis, J. (2012a) Building a New Governance Model for Environmental Sustainability, London, Stakeholder Forum. Available online at: www.stakeholderforum.org/fileadmin/files/Building%20a%20New%20Governance%20Model%20for%20UNEP%20AB.pdf

Laguna-Celis, J. (2012b). Ideas for a Sustainable Development Outlook, Earth System Governance Working Paper no. 24, Lund and Amsterdam, Earth System Governance Project. Available online at: www.ieg.earthsystemgovernance.org/sites/default/files/files/news/ESG-WorkingPaper-24_Lauguna-Celis.pdf

SDSN (2013) *An Action Agenda for Sustainable Development*, New York, Sustainable Development Solutions Network. Available online at: http://unsdsn.org/files/2013/06/130613-SDSN-An-Action-Agenda-for-Sustainable-Development-FINAL.pdf

Strong, M. (1998) A People's Earth Charter, Toronto, Earth Council. Available online at: www.earthcharterinaction.org/invent/images/uploads/Maurice%20Strong%20on%20A%20Peoples%20Earth%20Charter.pdf

UN DESA (2013) *Strengthening Public Participation at the United Nations for Sustainable Development: Dialogue, Debate, Dissent, Deliberation*, New York, UN. Available online at: www.globalpolicy.org/component/content/article/245-ngo-and-the-un/52458-strengthening-public-participation-at-the-un-for-sustainable-development.html

UNEP (2009) Second meeting of the Consultative Group of Ministers or High-Level Representatives on International Environmental Governance, Rome, 28–29 October 2009, Nairobi, UNEP. Available online at: www.unep.org/environmentalgovernance/Portals/8/documents/IEG-Draft-Report-Rome-final-edited.pdf

UNEP (2012) *The Fifth Global Environment Outlook*, Nairobi, UNEP. Available online at: www.unep.org/publications/contents/pub_details_search.asp?ID=6258

United Nations (2002) Report of the World Summit on Sustainable Development, Johannesburg, South Africa, 26 August–4 September 2002 (United Nations publication, Sales no. E.03.II.A.1 and corrigendum), chap. I, resolution 2, annex, para. 140 (d).

United Nations (2010) Keeping the Promise: United to Achieve the Millennium Development Goals. Available online at: www.un.org/en/mdg/summit2010/pdf/mdg%20outcome%20document.pdf

United Nations (2012a) Secretary-General's Remarks to Rio+20 Conference Ceremonial Opening Session, New York, UN. Available online at: www.un.org/sg/statements/index.asp?nid=6130

United Nations (2012b) *Realizing the Future We Want for All*, New York, UN. Available online at: www.un.org/millenniumgoals/pdf/Post_2015_UNTTreport.pdf

United Nations (2012c) National submissions for further review of the implementation of Resolution 61/16 on the strengthening of the Economic and Social Council, New York, UN. Available online at: www.un.org/en/ecosoc/about/strengthening ofecosoc.shtml

United Nations (2012d) United Nations Secretary-General Announces New Sustainable Development Initiative. Available online at: http://us2.campaign-archive2.com/?u=33cf89da7ade3a85156c5eda4&id=b2e297ce4d&e=69fb50bd2f

United Nations (2013) *A New Global Partnership: Eradicate Poverty and Transform Economies through Sustainable Development*, New York, UN. Available online at: www.un.org/en/development/desa/policy/untaskteam_undf/HLP%20P2015%20Report.pdf

Vidal, J. (2012) Rio+20: Reasons to be Cheerful, *Guardian* (London), June 27. Available online at: www.guardian.co.uk/global-development/poverty-matters/2012/jun/27/rio20-reasons-cheerful

Appendix I

# Stakeholder Forum

Stakeholder Forum is an NGO based in London and New York that has been working to advance sustainable development and promote democracy at a global level since 1987. The organization aims to enhance open, accountable and participatory decision-making on sustainable development by enhancing the involvement of stakeholders in intergovernmental processes. To this end the organization works with a diversity of stakeholders globally on international policy development and advocacy, stakeholder engagement and consultation, media and communications, and capacity-building – all with the ultimate objective of promoting progressive outcomes on sustainable development through a transparent and inclusive approach.

For the 2012 United Nations Conference on Sustainable Development (Rio+20), Stakeholder Forum engaged in a wide range of activities, including:

- convening regional capacity-building workshops to help stakeholders engage in the conference;
- undertaking stakeholder engagement activities such as green economy dialogues, and inputting recommendations from the sustainable development governance network;
- successfully engaging in policy and advocacy work around the agenda for improved corporate sustainability reporting and accountability;
- managing a wide range of media and communications projects, including a dedicated Rio+20 website for stakeholders.

As a follow-up to the conference, the organization is working with UN DESA and other actors to assist in the facilitation of stakeholder engagement around the intergovernmental process on Sustainable Development Goals (SDGs) and the High-Level Political Forum.

To complement its programme of work, Stakeholder Forum produces *Outreach*, a multi-stakeholder magazine published at key international sustainable development meetings. The publication has been produced since 1997 and continues to act as a conduit for stakeholders to share their opinions on the topics being negotiated at high-level meetings such as the United Nations Commission on Sustainable Development (CSD), the United Nations Governing Council and the United Nations Framework Convention on Climate Change (UNFCCC) Conference of the Parties (COP).

In addition, the organization is responsible for administering the Global Network of National Councils for Sustainable Development (NCSDs). The Global Network aims to help strengthen and enhance the work of NCSDs, seeing them better placed to meet their primary objective of furthering sustainable development at the national level.

Stakeholder Forum is also a leading author on sustainable development issues. Over the past 16 years the organization has published 14 books. The most recent include *Only One Earth: The Long Road via Rio to Sustainable Development* (2012), *Biodiversity and Ecosystem Insecurity: A Planet in Peril* (2011), *Energy and Climate Insecurity* (2009) and *Negotiating and Implementing Multilateral Environmental Agreements* (2007).

Appendix 2

# *Earth Negotiations Bulletin*

The *Earth Negotiations Bulletin* began as the joint initiative of three individuals from the NGO community who were participating in the preparations for the United Nations Conference on Environment and Development (UNCED). Johannah Bernstein, an environmental lawyer and director of the Canadian Participatory Committee for UNCED; Pamela Chasek, a doctoral student at Johns Hopkins School of Advanced International Studies; and Langston James Goree VI "Kimo," a former UNDP programme officer and NGO activist from the western Amazon, created the *Earth Summit Bulletin* in March 1992.

After publishing daily issues during the five weeks of the Fourth Preparatory Committee meeting for UNCED, the three raised funds to publish at the conference in Rio. Following the conclusion of UNCED, the International Institute for Sustainable Development (IISD) approached the three founders with an offer to continue publishing the *Earth Summit Bulletin* at follow-up negotiations to the Earth Summit. In November 1992 the *Earth Summit Bulletin* was renamed the *Earth Negotiations Bulletin*.

IISD Reporting Services has produced thousands of reports from hundreds of negotiations covering dozens of major multilateral environmental agreements. Its timely, up-to-the-minute reports and comprehensive online archives are widely respected and relied upon by experts and officials from around the world.

The IISD played a key role in providing a consistent stream of information on the entire Rio+20 process, from the early preparatory phase through to the conference in Rio. Working together with the UN DESA Secretariat, the IISD published daily and summary issues of its *Earth Negotiations Bulletin* from the first session of the CSD PrepCom in May 2010 through all of the "informal-informal" negotiations on the Zero Draft versions of the Outcome Document. The IISD also published *Earth Negotiations Bulletin* reports from each of the regional consultations held in conjunction with the UN Regional Economic Commissions. And in Rio during June 2012 the IISD published reports from five separate simultaneous events: (1) the third PrepCom and the UNCSD; (2) selected side events taking place during the conference; (3) the Rio Conventions Pavilion; (4) the Sustainable Development Dialogue Days; and (5) the World Congress on Justice, Governance and Law for Environmental Sustainability. These reports were distributed to the bulletin's more than 75,000 readers worldwide and throughout the Riocentro conference site in electronic format using the "ENB Mobile" handheld application for Android, iOS and HTML5 devices. The collection of IISD reports from the UNCSD can be found at www.iisd.ca/process/sustdevt.htm#uncsd

In addition to reporting from more than 25 UNCSD-related events in during two years, the IISD also published twice-weekly issues of the *Sustainable Development Update*, with original content stories of activities being undertaken by international organizations, governments and major groups. These newsletters were distributed on the UNCSD-L list, set up by the IISD as a community listserv and used by hundreds of organizations to distribute information on their own activities. A complete knowledgebase of Rio+20-related activities can be found on the Sustainable Development Policy and Practice website (http://uncsd.iisd.org).

# Index

The following abbreviations have been used in the index:

| | |
|---|---|
| EU | European Union |
| IFSD | Institutional Framework on Sustainable Development |
| MDGs | Millennium Development Goals |
| SDGs | Sustainable Development Goals |
| SIDS | Small Island Developing States |